ALEXANDER BRIDPORT BECHER

NAVIGATION
OF THE
ATLANTIC AND INDIAN OCEANS
AND THE
CHINA AND AUSTRALIAN SEAS

WITH AN ACCOUNT OF THE
WINDS, WEATHER, AND CURRENTS

FOUND THEREIN THROUGHOUT THE YEAR
ACCORDING TO THE MOST APPROVED AUTHORITIES,
INCLUDING EXTENSIVE EXTRACTS
FROM THE NAUTICAL MAGAZINE

Elibron Classics
www.elibron.com

Elibron Classics series.

ISBN 1-4021-6913-2 (paperback)
ISBN 1-4021-1938-0 (hardcover)

This Elibron Classics Replica Edition is an unabridged facsimile of the edition published in 1859 by J. D. Potter, London.

NAVIGATION

OF THE

ATLANTIC AND INDIAN OCEANS

AND THE

CHINA AND AUSTRALIAN SEAS

WITH AN ACCOUNT OF THE

WINDS, WEATHER, AND CURRENTS

FOUND THEREIN THROUGHOUT THE YEAR

ACCORDING TO THE MOST APPROVED AUTHORITIES, INCLUDING EXTENSIVE EXTRACTS FROM THE NAUTICAL MAGAZINE

Second Edition

WITH CHARTS

That seamen may with steam or sail
Know where to meet the favoring gale;
May take instruction from the skies,
And find the path where swiftness lies.

A. B. BECHER, CAPTAIN, R.N.,

Of the Hydrographic Office, Admiralty.

Author of the Voyage of H.M.S. Chanticleer—The Landfall of Columbus—The Storm Compass—Editor and Author of various Papers in the Nautical Magazine.

London

J. D. POTTER

31, POULTRY, AND 11, KING STREET, TOWER HILL.

1859

INTRODUCTION.

THE substance of the following pages is gathered from the valuable work of Capt. KERHALLET, compiled from the voyages of the most celebrated navigators, including the more recent contributions of our mercantile commanders to the "NAUTICAL MAGAZINE." It has already appeared in that periodical, from which it has gone through a first edition. The ready manner in which that edition has been taken up, affords good promise that the present, in its still more attractive form, will prove even more acceptable.

The subjects which it embraces are of much importance to the navigator; for, next to a fair wind, to know the season at the place to which he may be bound, how to go there, and when to go with regard to weather, is most desirable:—And such is the object briefly treated in these pages.

THE

ATLANTIC OCEAN.

GENERAL WINDS.

THE basin of the Atlantic Ocean, divided unequally between the West coasts of Europe and Africa and the East coast of America, presents a deep valley which appears bounded on the North and South only by the poles. Fields of ice have arrested the progress of navigators who have endeavoured to explore those regions.

In order to consider the winds common to this ocean, we shall divide it into three regions: the first, that comprised between the parallels of 30° North and South latitudes; the second between the latitude of 30° South and the South pole; the third between 30° North and the North pole. We shall divide each of these two last-named regions into two zones: the temperate zone, extending from the parallel of 30° to 60°, and the frozen zone between the latitude of 60° and the poles.

We shall first consider the winds of each of these divisions of the ocean at a distance from the coast; and then, commencing at one extremity, shall describe the winds generally met with near the shore.

Origin and Causes of the Winds.—The winds owe their origin to all actions which disturb the equilibrium of the atmosphere. Science has endeavoured to determine the cause of wind and the general laws by which it is governed. Philosophers attribute winds to the heat of the sun, different and variable on the surface of the earth, in conjunction with the diurnal motion of the earth itself.

The differences between the temperature of the polar regions and of those near the equator being very great, they assert that there must, of necessity, be a constant change of air between these regions. Thus the cold and dense air of the polar regions tends to replace the warm and expanded air of the equatorial regions, which latter, rising and forming a higher current, should transfer itself towards the North and South, in order to restore the equilibrium.

If the earth were still, the winds on its surface would generally blow from North and South, according to the hemisphere in which they are: but the earth, turning from West to East on its axis, with a quickness which increases in proportion as the equator is near; the consequence is, that in passing from high latitudes towards that great circle, currents of cold air arrive progressively in those regions where the rotatory motion is more considerable. Not being able to participate in this movement, on account of their want of cohesion with the earth, the winds take an opposite or inverted direction from that of the rotation of the globe, namely, from East to West.

Thus, by combining the rotatory motion of the earth, and the different temperatures of the surface, the currents of air, coming from the North and South in each hemisphere, are limited in their direction, and incline towards the N.E. and S.E., producing the winds called the general or trade winds of the torrid zone.

Without entering into the numerous theories on this subject, and the various objections to each, we shall confine ourselves to a statement of the leading authenticated facts regarding the winds of the Atlantic Ocean; the knowledge of which is of such vast importance in general to navigation.

North and South Polar Winds.—The two currents of air, blowing from the North and South poles towards the equator, abovementioned, are termed polar winds, North or South, according to the hemisphere in which they are found.

North and South Tropical Winds.—On the contrary, those are called North and South tropical winds, which, from the equator, are directed towards the poles. These last appear to be counter-currents to the polar winds.

First Division of the Atlantic or Torrid Zone.—In the northern hemisphere, the polar winds blow from the N.E.; in the southern hemisphere, from the S.E.; and they take a more easterly direction in proportion as they approach the equator. Between the tropics, these winds have received the more common name of "*trade winds* of each hemisphere." They are also called the "general winds of the torrid zone." These winds render the passage across the Atlantic, from the Old to the New World, quick and easy.

The trade winds appear to be the only primitive winds. Where they are established, the weather is always fine, and the sky generally clear. If they cease for a while, the sky becomes clouded, and, in certain parts, storms are experienced, the more lasting and severe in proportion as the places are more or less distant from the equator.

Those regions where the trade winds do not prevail, are constantly exposed to squally and tempestuous weather; where they cease only from any cause, bad weather is experienced, and it has been remarked that they always return with some violent reaction, or with torrents of rain.

The trade winds to the North and South of the equator have similar characters, as will be hereafter described.

Limits of the Trade Winds.—The polar limits of the trade winds from N.E. and S.E. generally extend on each side of the equator to the parallels of 30° North and South latitude. Nevertheless, this limitation differs greatly in some parts of the ocean; because it is influenced by temperature, and varies about 3° North or South, according as the declination of the sun is North or South.

The equatorial limit of the N.E. and S.E. trade winds is generally variable from the same causes. That of the N.E. trade is about the mean parallel of 8° North latitude; that of the S.E. trade is at the parallel of 2° or 3° North latitude.

From a numerous collection of observations, the following table has been formed; which, however, only affords an approximation to these limits:—

Periods of the Year.	Polar Limit of N.E. and S.E. Trade Winds.	Equatorial Limit of N.E. Trade Wind.	Equatorial Limit of S.E. Trade Wind.	Polar Limit of N.E. Trade Wind, according to the Months.
	° ′	° ′	° ′	° ′
Winter. ..	24 45 N. or S.	5 45 N.	2 30 N.	January . 23 24 N.
				February. 28 39
				March .. 27 19
Spring....	28 0	5 45	1 30	April ... 28 18
				May..... 28 31
				June 31 25
Summer..	30 45	11 20	3 15	July 29 36
				August .. 31 11
				September 32 4
Autumn ..	28 20	10 0	3 15	October.. 25 38
				November 27 14
				December 22 15

In the Atlantic Ocean, the trade wind from the N.E. in the part comprised between Cape Verd and the coast of Guinea, has less force and constancy than that from the S.E. in the neighbourhood of the equator; owing, doubtless, to the form of the coasts which bound the ocean in this part. Nevertheless, in the neighbourhood of the Antilles, the trade wind generally blows strongly, varying from East to N.E.

Peculiarities of the Trade Winds.—It has been observed, that in the zone comprised between the equator and the parallels of 28° North and South latitude, in proportion as the sun approaches the equator, the winds blow, in the northern hemisphere almost from N.E., and in the southern hemisphere from S.E.

But if the sun is in the northern hemisphere, and at its greatest distance from the equator, the winds in that hemisphere have a tendency to blow more from the East, and more violent storms are then experienced than at any other time. In the southern hemisphere, the trade wind then blows more from the South.

If, on the contrary, the sun is in the southern hemisphere, the same facts are produced in an inverted order; thus, in this hemisphere the wind blows more from the East, while in the northern hemisphere it veers towards the North, and in this case they reach their nearest approach to the equator. And generally, in either zone, rain, sudden gusts of wind, and storms must be expected in those places where the sun is vertical.

During winter, the northern trade is sometimes met before reaching the latitude of Madeira. This fact is, however, only an exception to the general rule above laid down in alluding to the polar limit of this wind. At other times, the variable winds of the temperate zone extend as far as 20° North, without appearing subject to any general law; and this, too, in all seasons of the year.

In the southern hemisphere similar facts appear. Thus, during the fine season, the southern limit of the trade winds from the S.E. is found to be about the parallel of the Cape of Good Hope; while, from June to August, westerly winds are prevalent between that parallel and the tropic of Capricorn.

It has been also remarked, that near the polar limit of the trade winds, calms and light variable winds are often met with, producing rain, and this in a very extended zone, namely, that which separates the tropics from the parallels of 29° North and South.

Thus we see that the polar limits of the trade winds are very variable. In the southern hemisphere this limit is sometimes near the tropic, but more frequently on the parallel of the Cape of Good Hope.

In this same zone it has been observed, with regard to the S.E. trade wind, that in latitudes South of 16° S., the wind has a greater tendency to blow from the N.E. than from the S.E., so as to blow rather from East to N.E. than from East to S.E. This variation is also pretty frequent. During a year's stay at St. Helena, Halley found that, in that island, the trade winds always blew from S.E. or nearly so; and that they more frequently veered from S.E. to East than from S.E. to South. During the East wind the weather was gloomy, and the return of fine weather depended on the wind from the S.E.

And, lastly, it may be stated that in approaching the coast of America the polar limits of the N.E. and S.E. trade winds extend some three or four degrees more towards the North in the northern hemisphere, and more to the South in the southern hemisphere than on the coast of Africa.

Deviations in the Trade Winds.—It happens sometimes in the region of the trade winds, that winds from opposite directions interrupt their usual course.* These

* Columbus, who was the first discoverer of the trade wind, in 1492, was the first to discover this irregularity; and it is remarkable that it was of service to him in quieting the fears of his crew, who, having observed the constancy of the wind from the eastward, believed that they never would be able to get back to Spain. This irregular wind broke the spell, and much to the satisfaction of Columbus, who was then beginning to feel the inconvenience of that mutinous spirit which, in a later part of the voyage, had nearly cost him his life.—*Landfall of Columbus, by A. B. Becher, Captain R.N., published by Potter, 31, Poultry, London.*

winds are never of long duration, and only arise from accidental causes.

In the neighbourhood of the islands situated in the zone of the trade wind, this wind is also interrupted. Thus, among the Cape Verd Islands, the N.E. trade wind is often lost; and in the zone comprised between the parallel of 10° latitude and the equator, and also from the meridian of Cape Verd to the most westerly meridian of this Archipelago, it is observed that there is, in reality, no settled wind, but only breezes of short duration.

In the vicinity of Trinidad, situated near the coast of Brazil, frequent changes have been observed in the trade wind from S.E. to South of these islands, coming more generally from the northward than from the southward, as well as sudden gusts from the West. In the two last-mentioned cases, the vicinity of the coasts of Africa and America doubtless occasions these disturbances.

Variable Winds of the Torrid Zone.—The South-eastern trade wind is separated on its equatorial border by a zone of calms and changeable winds, varying considerably in extent from North to South, the mean latitude of which is about 8° North. In this zone, calms, squalls, rain, and light breezes, blowing from all points of the horizon, are met with and chiefly from S.W. This zone, during summer, reaches further North of the equator, and is then found sometimes as far as the parallel of 14° and even 15° North latitude. In winter, on the contrary, it does not reach so far; and when the sun enters the tropic of Capricorn, it is not found beyond

3° N. latitude. It always keeps, however, North of the equator.

The following table is drawn up by Horsburgh, from observations made between 1791 and 1807, contributed by the journals of two hundred and fifty ships, which have crossed the zone of the variable winds of the equator, between 30° and 40° West longitude.

Month.	Sailing South the N.E. Trade Wind is lost in North lat. from	Returning North the N.E. Trade Wind is found in North lat. from	Mean.	Returning North the S.E. Trade Wind is lost in Latitude from	Sailing South the S.E. Trade Wind is found in North Lat. from	Mean.	Lat. of the Zone of Variable Winds of the Torrid Zone.
	° °	° °	° ′	° °	° °	° ′	° ′
January.	5 to 10	3 to 6	5 45	0½ to 4 N.	2 to 4	2 45	3 0
February	5 .. 10	2 .. 7	6 0	2 S... 3	0½.. 1	1 15	4 45
March ..	2½.. 8	2 .. 7	5 0	1 .. 2	0½.. 2½	1 1ç	3 45
April ...	4 .. 9	4 .. 8	5 45	2 .. 2½	0 .. 2½	1 15	4 30
May....	5 .. 10	4½.. 7	6 30	1 N... 4	0 .. 4	2 45	3 45
June ...	7 .. 13	7 .. 12	9 0	1 .. 5	0 .. 5	3 0	6 0
July ...	8½.. 15	11 .. 14	12 0	1 .. 6	1 .. 5	3 30	8 30
August .	11 .. 15	11 .. 14½	13 0	3 .. 5	1 .. 3	3 15	9 45
Septemb.	9 .. 14	11 .. 14	14 45	2 .. 4	1 .. 3	3 0	8 45
October.	7½ . 13	8½.. 14	10 0	2 .. 5	1 .. 5	3 0	7 0
Novemb.	6 .. 11	7 ., 10	8 0	3 .. 4	3 .. 5	3 45	4 15
Decemb.	5 .. 7	3 .. 6	5 30	1 .. 4	1 .. 4½	3 55	2 30

Favourable Time for crossing the Line.—The foregoing table shows the zone of the variable winds to be larger from June to December, and less from December to June,—an important condition in making a voyage, because, as a ship must go from one hemisphere to the other, and cross the equator, thus passing from the North or South, the most favourable months for crossing the zone of the variables will be those from December to June. According to the time of crossing the line the passage will be several weeks longer or shorter, as it is more or less interrupted by calms, squalls, or variable winds near the equator. And we must also expect it to be longest from these causes during the months of June, July, August, and September.

It may also be seen that, in the northern hemisphere, the N.E. trade wind verges, according to the season, more or less towards the equator; but it rarely passes to the southward of it. On the contrary, the S.E. trade wind reaches it easily, and northward sometimes even so far as 5° North latitude, and that in the neighbourhood of the coast of America.

Union of the Trades.—It happens, however, that the N.E. and S.E. trade winds join each other generally somewhere about the meridian of 28° or 33° West longitude, where a ship may pass perhaps in a squall from one of these winds to the other without experiencing any calm. Sometimes even in the vicinity of the African coast, and generally from December to February, the N.E. winds nearly join those from S.E. The different directions of the trade winds from N.E. and S.E. occasion great uncertainty about the weather and the winds within their limits; an uncertainty which increases on approaching the zone of either. Besides this, it has been observed that near the equator the winds change more frequently from East to South than from East to North. We find, nevertheless, in the region of the variables of the equator, winds which blow from West to N.W., and from West to South, and generally from July to September.

The preceding considerations are of great importance in the general navigation of the Atlantic Ocean, and more particularly in that of the gulf of Guinea. We shall apply them hereafter when alluding to the routes for crossing the Atlantic; for, according to that which may be adopted, the place of crossing the equator becomes important.

Extent of the Trade Winds.—The preceding observations on the winds of the first portion of the Atlantic Ocean refer to a considerable distance from the coast, which has some effect in modifying their direction as well as their force. It is observed that the North and South trade winds sometimes reach as far as the coasts of America, though they are not found regular till about 140 leagues from the African coast, and about 160 leagues from that part of it lying between Cape Palmas and Cape Verd. In this latitude, occasionally between June and September, and even till October, the variables are found even in 28° West longitude, blowing from West to N.W. and S.W., interrupted by calms, accompanied by rain and bad weather. The islands in this region also influence the trade wind in their vicinity, but much less sensibly than the continent. The higher these islands are, the more frequent is rain about them, at least so observation has shown; and it is generally found in the torrid zone that the wind which is constant at sea, becomes changeable near elevated and extensive coasts.

Seasons of the Torrid Zone.—In the Atlantic Ocean, within the torrid zone, the weather is very variable, according to the latitude and time of year. It is observed that the atmosphere is more unsettled during the spring and autumn months, generally at the time of the equinoxes, than during the summer and winter months.

In this zone, North of the equator, the rainy season begins when it is quite dry to the South of it, and *vice versâ.* And it lasts from April or May till September or October. The dry season begins in October or November and lasts till April or May. South of the equator the weather and seasons, in this zone, are contrary to those of the region North of the equator, and they change nearly in the same months.

As a general rule, it may be assumed that in both hemispheres the rainy or winter season commences at any place when the sun, moving from the equator, passes the zenith of that place; and the dry season begins when the sun, returning towards the equator, has repassed the zenith of the same place.

Clearly as this law indicates the seasons, it is nevertheless subject to many exceptions. Thus, the seasons do not begin and finish precisely at the period of these transits, but according to the time at which they take place. There is also always a period of doubtful uncertain weather, lasting a longer or shorter time, between each season, with variable winds, calms, and squalls, some of which have received the name of *tornadoes.*

In several places these squalls occur at the beginning and end of the rainy season; and limit the season of the great rains.

According to the foregoing law, it may be seen that the duration of the winter season at a place depends greatly on its latitude, and that it should be proportionably lengthened as it is nearer to the equator.

The hottest season within the torrid zone is that of winter, which is also the time of variables and calms. In the fine or dry season, on the contrary, the breezes are fresh and regular; this is generally the period when, near the coasts, land and sea breezes prevail with regularity.

At places within the immediate vicinity of the equator, four seasons have been distinguished, two dry and two rainy; but, in reality, that part of the rainy season when rain is less abundant, and when rain squalls are separated by intervals of tolerable weather, has been considered one of the dry seasons.

Second Part of the Atlantic Ocean.—The second region of the Atlantic, as above said, is that comprised between the parallel of 30° South and the South pole; in which two zones are distinguished, the frigid and temperate. Let us now allude to the first.

Frigid Zone.—In the frigid zone, where but few navigators have penetrated, observations are necessarily few, and these relate only to one season, namely, the summer; that being the only time when it can be penetrated to the South. The celebrated Cook, Admiral Dumont D'Urville, and Sir James Ross, are the principal navigators who have penetrated into this dreary region of the Atlantic; Cook proved that between 60° and 70° S., the winds are generally moderate; Foster adds that they often blow from the East. Cook has also observed, that in those high latitudes the currents, though not strong, drift the ice towards the N.E., the North, and the N.W. D'Urville, after a stay of forty-nine days in that region, in which he could not penetrate beyond the latitude of 63° 33′ S., and was blocked up by the ice in 62° 22′ S. and 39° W., has left us the following observations, which appear in the voyage of the *Astrolabe* of 1837.

Variable winds from the East, between N.N.E.. S.S.E.. and South for twenty-seven days; westerly winds, varying from N.N.W. to S.S.W. twenty-two days, during the months of January, February, and the beginning of March. During this time, the sloops *Astrolabe* and *Zelée* experienced a breeze from the North, accompanied by fog and rain; with winds from S.S.W. and S.W. the weather was alternately fine and cloudy; the winds from this quarter were in general moderate, and sometimes fresh.

The winds from N.E. and E.N.E. sometimes brought fine weather, but the sky was more frequently cloudy and foggy; snow fell principally with the winds from E.S.E. to N.N.W. by the N.E.; with the same breezes there was fog. Lastly, the strongest were those from East, E.S.E. and S.S.E., which in general blew very fresh.

Sir James Ross crossed the parallel of 60° S. latitude on the 23rd of December, 1842. During his stay in the frigid zone, which lasted till the 1st of March, he made observations on the winds, of which the following is a summary:—

December, 1842.—Winds from West changing from N.W. to S.W., six days; winds from North, half a day; winds from South, a day and a half; winds from S.E., one day, moderate; those from South blowing very hard.

January, 1843.—Easterly winds changing from N.N.E. to S.E., twelve days; northerly winds, two days; southerly winds, three days and a half; westerly winds, varying from N.N.W. to S.S.W., thirteen days and a half. In this month, two days of strong breezes from N.W., the other winds fresh or moderate.

February.—Easterly winds, changing from N.N.E. to S.S.E., eighteen days and a half; northerly winds, two days; South winds, three days; westerly winds, veering from N.W. to S.W., four days and a half. One day, a hard gale from the North; one day, a stiff breeze from the East; the rest fresh or moderate.

From the preceding observations it appears that during summer in the frigid zone of the second region of the Atlantic Ocean, the East and West winds are tolerably equal, but that the East winds rather prevail.

Temperate Zone.—In the temperate zone of this same region, westerly winds, extending as far as the tropic of Capricorn, and sometimes even to the latitude of 20° S., are predominant, varying from N.W. to S.W.; but the winds in this part of the Atlantic Ocean are changeable and irregular.

It has been remarked that in a zone comprised between the parallel of 28° and 35° S. latitude, the winds are extremely variable; but those which are most frequently met with are from N.E. to N.W.b.N. and from N.W. to S.W.b.W., principally during June, July, and August.*

The prevailing westerly winds, varying from N.W. to S.W., between the parallels of 30° and 50° S., appear to be produced by the meeting of the returning currents from West to East, occasioned by the trade winds, (and named here tropical winds,) with the polar currents proceeding from the poles towards the equator. The winds

* Dampier.

which result from these two currents of air ought to take almost a mean course, depending on the relative force of the two currents, and, consequently, this direction will be very variable, although it is generally from the West.

It has been observed that between 30° and 50° S. latitude, the winds blow periodically from S.W. to N.W.; thus, they vary from West to N.W. while the sun has South declination, and during the rest of the year they are in general from West to S.W., attended by bad weather.* In this zone the easterly winds are never of long duration. When the winds shift to the South a calm soon takes place; and it has been observed that between Cape Horn and the Cape of Good Hope, when the wind continues from the North for several days in succession, it brings dark rainy weather; if these winds shift to South of West, the weather clears up and becomes fine.†

Third Region of the Atlantic.—The third region of the Atlantic is comprised between the parallel of 30° N. latitude and the North pole, and is here divided, like the preceding one, into two zones—the frigid and temperate.

Frigid Zone.—In the frigid zone, comprised between 60° N. latitude and the pole, and between Europe and America, and which includes Spitzbergen and Iceland, the winds are variable. The snow-clad lands in this zone, and the ice which prevails there, exercise a most powerful influence on the winds, varying them according to the seasons. The different navigators who have tra-

* Nicholson. † D'Apres di Mannevillette.

versed this zone, in what is called the northern frozen ocean, have not observed any law in the winds,* principally during summer.† However, several navigators agree in acknowledging the northerly winds as regular and prevalent.

Every wind in this zone is accompanied by a low temperature and snow, excepting in July and August, and a part of June. At this period, with southerly winds, the weather is tolerably fine, although attended by snow and rain. The same winds bring fog during those months. The coldest winds are those from North and N.E.; during June and July, winds from S.S.W. are frequently met with, sometimes blowing with much violence. In April and May, South winds are attended by snow; in the rest of the year there are thick fogs, and very bad weather.

Spitzbergen.—At Spitzbergen it is said that during the first months of the year southerly winds prevail, and during the other months those from the North; but the S.E. and N.E. winds are those which bring most snow. Annexed are the observations of Sir Edward Parry, during the months of May, June, July, and August, 1827, on the winds of those regions comprised between 70° and 82° 40′ N. latitude, as given in the narrative of his voyage towards the North Pole in the *Hecla*.

North	7½	days.	W.S.W.	1	days.
N.N.W.	5½	,,	S.W.	9½	,,
N.W.............	9	,,	S.S.W.	1	,,
W.N.W.	2	,,	South	4	,,
West	13	,,	S.S.E.	2	,,

* Phipps. † Standidge.

S.E.	12	days	N.E.	9½	days.
E.S.E.	½	"	N.N.E.	11	"
East	17½	"	Calms	13½	"
E.N.E.			Variables	4½	"

Nova Zembla.—At Nova Zembla, from September to May northerly winds prevail almost without interruption; and from May till August those from West.

Temperate Zone.—In the temperate zone, comprised between the latitudes 30° and 60° N., the prevailing winds are, like those of the corresponding zone of the other hemisphere, from the West, varying from N.W. to S.W. The prevailing West winds are doubtless produced by the same causes as those already stated in alluding to the corresponding zone of the southern hemisphere. They are produced by those currents of air, named "tropical winds," blowing from West to East, and the polar currents directed from North to South. Their direction, therefore, depends on the relative intensity of these currents, and might be a middle one.

There is an important fact which shows the prevalence of S.W. winds in the temperate zone of the northern hemisphere, consisting in the difference of time occupied by the passage from the North of Europe to North America and that of the return passage home from North America to Europe. From Liverpool to New York the voyage is at least forty days; while from New York to Liverpool it is only about twenty-three.

The mean direction from whence is the prevailing wind of this zone, has been determined by Kramtz from numerous observations as follows:—France S. 88° W.; England S. 66° W.; Germany S. 76° W.; Denmark

S. 62° W.; Sweden S. 50° W.; Russia N. 87° W.; North America S. 86° W.: showing that Russia is the only country where the wind has a little North tendency in its source.

In the Atlantic Ocean the direction of the prevailing wind is generally from S. 45° W. to S. 70° W.

When the sun is in the northern hemisphere the prevailing westerly winds are S.W. and W.S.W.; if, on the contrary, the sun is in the southern hemisphere, they are from W.N.W. and N.W. This last period is that of gales and bad weather between North America and Europe.

Having described the winds met with *at sea* in the Atlantic Ocean, it remains for us now, in order to complete our view of the winds of this ocean, to allude to those which are experienced on its coasts. We shall commence with the coast of Africa, offering first some general remarks as to that coast.

Land and Sea Breezes.—The land and sea breezes blow on certain parts of it with great regularity, particularly on those of the intertropical continents, and the islands situated in the torrid zone, so much so that they might be reckoned in the periodical winds. In general the sea breeze blows during the day, and the land breeze commencing towards evening lasts part of the night, blowing in the opposite direction; it commonly ceases shortly after sunrise, and rarely lasts beyond nine or ten o'clock in the morning, Philosophers attribute these breezes to the difference of temperature between the sea and the land.

It is observed that in the neighbourhood of the African

coast of Morocco, the sea breeze comes most frequently from N.W.; on the coast of Guinea from South to S.W., and on that of Loango and Congo from S.W. to West. These breezes generally blow alternately with those from the land. They are generally found as stated near the coast, but they sometimes prevail far from it, gradually losing their force, until they are opposed and overcome by the general winds. The parts where these changes occur, are almost always subject to storms, accompanied by rain or calm.

The following general view is given by philosophers of the alternate land and sea breezes. In the morning, the temperature of the land and sea being nearly alike, the air is at rest. But when the sun rises above the horizon and the earth begins to be more heated than the water, the sea breeze begins to be felt; gentle at first, but gradually gaining force, till it attains its height with the greatest heat of the day. It then diminishes gradually with the decrease of temperature till night; when an interval of calm takes place. Again, during the night, when the land is colder than the sea, the land breeze rises, and attains its greatest strength at the time of the highest temperature of the night, It continues until day, still diminishing, and sometimes even until eight or nine o'clock in the morning, according to the latitudes. A knowledge of these phenomena is highly useful, and is particularly advantageous in local navigation, when it may be turned to good account in shortening passages.

Solar Breezes.—There are yet, on some parts of the western coast of Africa, certain winds, which may be called solar, their changes appearing to be regulated by the influence of the sun. We find them principally in the North, on the coast of Senegambia; and also in that extending from Cape Lopez to Cape Negro, to the South of the equator.

On the former coast these winds vary from N.E. to N.W.; on the second, from S.E. and South to S.W. and even to W.S.W. The course of these winds, which blow strong on the coast of Senegambia, principally from November to April; and on the other from October till March, and even April, is as follows:—The breeze at sea lasts through the day, strengthening in the afternoon, then weakening towards the evening; during the night, it changes, and blows more along the land.

In navigating these two coasts, a ship should regulate her sailing so as to be near the land when the night breeze springs up, and out at sea when that of the day begins. These winds, the variations of which occasion the alternate land and sea breezes, do not, however, blow directly off the shore. They are sometimes very strong, though at a short distance from the coast; and in order to profit by them it is necessary to make short boards of twelve, fifteen, or thirty miles at farthest from it.

The Harmattan.—The harmattan is a wind peculiar to the western coast of Africa; it blows from East, changing to E.N.E. on the North coast, from the parallel of Madeira to that of the River Gabon. It is sometimes very strong on the coasts of Senegal and Senegambia, but often light on the coast of Guinea. It is a cold dry wind, generally lasting for a series of three, six, or nine days, principally between the end of November and

February or March. It reaches, however, but a short distance from the coast, sometimes commencing at sunrise and ending in the afternoon. This wind is often accompanied by fog, and sometimes loaded with a fine reddish dust, so thickly that nothing can be distinguished, particularly near the coast, at a trifling distance.

At Senegal this is an extraordinary dry wind, and produces the most singular effects. It is said to be generally healthy, though certainly very disagreeable and inconvenient. The greatest advantage derived from it, nevertheless, is the production of gum. It suddenly arrests the circulation of sap in trees that is flowing rapidly at the season when the winter rains are just ended, and causes it to exude from the wood, thus forming gum, constituting the principal commercial greatness of Senegal.

Tornadoes.—Tornadoes are violent squalls of short duration, but are very frequent on the West coast of Northern Africa. In the South they are seldom felt, and are always slight. The name is derived from the sudden changing or veering of the wind while they last. Their approach is made known long beforehand by clouds of a dull yellow colour, which by night are very black. They originate to the North and N.E., and generally move against the prevailing wind. Tornadoes are generally distinguished by electric phenomena, and they gradually overcome the prevailing wind.

Every precaution becomes necessary to meet a tornado. Dark heavy clouds gradually spread on the horizon, and lifting slowly, leaving an extensive arch clear and distinct, are furrowed every now and then by lightning; the clearer the arch is defined the more violent will be the approaching storm.

A few moments of calm then ensue, and the N.E. wind is suddenly felt coming violently and rapidly with the storm, which breaks forth in all its fury when the arch attains the height of thirty or forty degrees above the horizon.

From N.E. the wind veers rapidly to East or S.E. blowing with the same force. The storm afterwards ceases, with rain, and when the wind slackens in force it veers to the South, and when it becomes still less to S.W. A calm frequently succeeds a tornado; and it is found that when the wind precedes the rain, the storm is more violent.

Vessels to meet these passing squalls, which never last more than an hour and a half, should shorten all sail, and furl, if possible. We never know what the extent of these violent commotions of the atmosphere may be, which sometimes have the force of hurricanes of short duration, in which the wind changes so suddenly that a sail might be split and inevitably lost.

The S.W. winds of the West coast of Africa are often disturbed at certain periods, (especially in the winter,) for we find in the Mediterranean, and on the coast of Portugal, intense polar currents, which, increasing the rapidity of the trade wind from North, forces it abruptly towards the equator. The meeting of these winds and those from S.W. prevailing at this period, contributes, perhaps, to produce tornadoes. They are very frequent in the northern hemisphere, and also very violent. South of the equator, on the contrary, as already observed, these disturbances of the atmosphere are rarely felt.

On the coast of Gaboon, and in the gulf of Biafra, tornadoes blowing from N.W. to West and S.W. sometimes though very rarely occur. In the northern hemisphere they principally take place at the commencement of the winter season, and consequently the time of their appearance varies according to the place. Thus, they are found at Cape Palmas a month earlier than at Sierra Leone; while at Goree and St. Louis, to the northward, they are a month and a half after they have prevailed at Sierra Leone. In several localities they take place exactly at the end of the winter season, principally from the archipelago of the Bissagos to Cape Palmas and the coast of Guinea. In the gulf of Guinea they come chiefly in the months of March, April, and May, and they again occur in some localities during November and December.

South of the equator the tornadoes blow chiefly from S.E., and take place from March to June, and from September to October. They diminish in force in proportion as they are found to the South; and in the latitude of the Kongo, it is not unusual for storms, without much wind, to form in the East, and passing South, to terminate in the S.W. They are much like the storms which are encountered in the northern hemisphere. These are somewhat similar to tornadoes in the changes of the wind, but not in force.

On the coasts of Angola and Benguela, towards the evenings of November and December, storms of this description are found, but very rarely accompanied by much wind. On the contrary, the clouds heap together, and the wind falls gradually to a calm. These clouds generally become scattered at ten or eleven o'clock at night, and at the same time a feeble land breeze rises, which usually lasts during the remainder of it.

Cape of Good Hope.—About two hundred leagues West of the Cape of Good Hope, S.E. winds prevail from October to March, and even April. In the same latitude, from May to August, at a hundred leagues to the West of the Cape, winds changing mostly from N.W. to S.W., are found, bringing a high sea with dirty weather. The same winds extend two hundred leagues to the East of the Cape of Good Hope, and the weather becomes worse as it is approached, during this season, from East to West.

N.W. winds here bring fog, rain, and haze; but the weather is fine and cold when it blows from S.W. During April and May these winds are found equally to the East and West of the Cape, but only as sudden gusts. These gusts are preceded by black clouds gathering in the West. The wind then begins blowing violently from W.N.W. to West, and then changes rapidly to S.W.; it then passes to the South, afterwards moderates, and soon after a calm ensues.

Doubling the Cape of Good Hope from Eastward.—When the season is advanced, D'Apres de Mannevilette advises a ship that would round the Cape from the East, not to go further than forty miles from land, and not nearer to it than eighteen miles, in order to preserve a latitude where the winds are less violent and the sea not so high as to the southward.

These stormy winds are very frequent during the winter. They are accompanied by so much rain, that sometimes two successive fine days can scarcely be found. According to several mariners, this bad weather is felt as far East as Madagascar.

On the parallel of 36° S. latitude, at 200 or 250 leagues from the Cape of Good Hope, to the East and West of the Cape, the N.W. winds, which are so violent near it, become moderate, and frequently vary to S.S.W. Generally between the parallels of 33° and 36° S. latitude, the West winds appear to prevail.

Doubling the Cape from Westward.—Consequently, when doubling the Cape from the West, it will be best to keep well off the land on the parallel of lat. 35° or 36° S., and to enter the Indian Ocean nearly on that of 36°. In adopting this course, a vessel would also profit by the *cross* current of the Atlantic Ocean setting eastward.

Winds at the Cape of Good Hope.—At the Cape of Good Hope and in Table Bay the months of September, October, and November form the spring; those of December, January, and February the summer; March, April, and May are the autumn; and June, July, and August the winter months. The following table, constructed from observations made in the course of many years at Cape Town, shows the prevalent winds during the year and in each month.

Months.	Winds.	Remarks.
January	S.E.	Dry, hot, occasional rain, with wind from N.W.
February	S.E.	Temperature variable, heavy rain occasionally with N.W. winds.
March	S.E.	Heavy gusts from N.W.; thunder; light rain and mist.
April	S.E. & N.W.	Heavy gusts; temperature variable, and mist.
May	N.W.	Fine at the beginning of the month; thunder and tempestuous at the end.
June	N.W.	Heavy gusts sometimes from S.E. or N.E.; rain, thunder, and stormy.
July	N. & N.N.W.	Frequent gusts of wind; cold, mist, snow, rain, hail.
August	N.W.	Ditto.
September	S.E.	Weather changeable and mild.
October	N.W.	Heavy rain; thunder and lightning.
November	N.W. & S.E	Hot dry weather; moderate breezes.
December	S.E.	When the wind blows from N.W. the breezes are light,—the weather hot and dry.

During winter, ships cannot anchor in Table Bay; and to refit at this season they anchor in False Bay. The approach of winter in the neighbourhood of the Cape of Good Hope is indicated by the prevailing S.E. winds being interrupted occasionally, and also lessening in force.

Trade Winds of the West Coast of Africa between the Cape of Good Hope and Cape Palmas.—If a line be drawn from the Cape of Good Hope to Cape Palmas, it will nearly be that of separation between the S.E. trade winds and those prevailing winds, varying from S.S.E. to S.S.W. and S.W., which blow during the whole year in that part of the ocean between the above line and the West coast of Africa. The distance from the coast at which these winds prevail is variable, as the limit itself is, and they are found much stronger on approaching the Cape. It has also been remarked that on this part of the African coast the wind frequently takes a direction making an angle of about two points with the line of the coast.

At eighty or a hundred leagues from the North coast of Guinea, and on the line of separation above-mentioned, the trade wind is generally found; which at this distance begins also to incline towards the coast, and in proportion as this distance is lessened draws to South and S.S.W. and even to S.W. On this limit of the trade wind calms, storms, and variable winds are generally found.

Hottentot Coast.—During nearly the whole year on the Hottentot coast, and that of Cumbasea, strong breezes from South prevail, changing from S.S.E. to S.S.W., sometimes in heavy gusts of wind. In proportion as the coast is left, these breezes diminish in strength, veering to South and S.S.E. in order to join the S.E. trade.

Coast of Benguela and Angola.—On the coast of Benguela and Angola the weather is generally fine all the year, except during the months of March and April. In November, December, January, and February, S.S.W. winds blow fresh; also S.W. and W.S.W., and now and then those from W.N.W.; so that in these months navigation is easy on this coast. In November and December there is sometimes a little small rain, especially in the mornings, with the wind from S.E. or at least South. Directly it draws to S.W. the weather clears off and the sky suddenly lightens up. Sometimes, again, there are appearances of stormy weather and lightning, particularly in the evenings; but, as already noticed, these storms bring but little wind, and this frequently falls away until it gradually becomes calm. These appearances seldom, as already stated, last longer than ten or eleven o'clock at night; and then a gentle land breeze rises, which generally lasts till morning. March and April are the two worst months of the year, on account of their stormy character. However, as soon as the wind veers to S.W. the sky clears and the weather becomes fine. The land breeze then comes on from S.E., and sometimes from N.E.; but these gusts are quite unlike the tornadoes North of the equator.

From May to October, during the "fine season," the sky is often overcast, especially in the morning, and the sea breeze is then seldom strong; while, on the contrary, it rarely fails when the sky becomes clear, and is still fresher when it clears quickly. In May and June the calms are less. The sea breeze begins late, and the land wind is fresh till after sunrise. In July, August, September, and October, S.W. winds are fresh and well established. They are found near the coast at ten or eleven o'clock in the morning; falling towards sunset, and rarely last later than seven or eight o'clock in the evening. During October they sometimes last till midnight, and are followed by the land breeze till eight or nine o'clock in the morning. In the interval between these breezes there is sometimes a calm.

Coast of Congo.—On the coast of Congo the breezes are generally moderate, and from September to March between South and West. From March to October the prevailing winds are from S.S.E., and are sometimes strong from between East and North. Strong sea breezes are sometimes found between North and West, generally from April to August. During this season there are very heavy rains. In the fine season, from September to March, the land and sea breezes succeed each other regularly. But they are not so regular during the rainy season, which lasts only three months, from November till February.

Rio Congo.—In the River Congo the seasons and winds are nearly the same as on the coast to the South of this river;—observing that the further South, the later are the seasons. Thus, in the River Congo the rainy season is from October to January.

Coast of Loango.—On the coast of Loango, from September to March, the prevailing winds are from South to West. In December and January strong winds prevail from West and S.W. From March to October the wind is generally from S.S.E., changing to South and S.S.W.

The alternate land and sea breezes are very regular on this coast, except during the rainy season—from September to December. Tornadoes take place in March, April, May, and often in September and October; sometimes also in January and February. These tornadoes, generally less violent than those North of the equator, are, however, sufficiently severe to oblige ships to reduce all sail.

Cape Lopez.—Off Cape Lopez, from June to October, the wind is generally from South, moderate, as well as that from S.S.W., which wind prevails during the other months.

At the end of November storms with heavy rains occur. The tornadoes are most severe in March and April. They also occur in November, December, and January;

as well as storms, which differ from tornadoes only by the wind being less violent.

Gulf of Biafra.—On the eastern shore of the Gulf of Biafra, two seasons only are generally known; that of tornadoes and bad weather lasting from March till the middle of September. July and August are comparatively the dry months, in which the S.S.W. breezes are fresh; they veer to the S.S.E. sometimes and blow fresh, commencing North of the equator as far as lat. 2° or 3°. The rainy season begins in September and lasts till March. This is the time of calms and light breezes from South to W.S.W.

The islands of the Gulf of Biafra—Princes Island, St. Thomas, and Anno Bom—have the same winds as the adjacent coast. The alternate land and sea breezes are more regular and fresh near them, and cease about the time of the rainy season. The former never reach far from the coast either of the continent or the islands.

North Coast of Biafra, Coast of Benin and St. Paul to Cape Palmas.—On the North coast of the Gulf of Biafra and Benin, and on the shore of the Gulf of Guinea, a moderate trade wind comes from S.W. to West, and prevails, with more or less regularity, according to the season, during the whole year. The harmattan on this coast, blows in November, December, and January, from the East, but never strong.

From October to February, the period called the fine season, alternate land and sea breezes are found near the coast. The former never extend further than four leagues off shore and are always weak. The tornadoes on this coast occur from March to May. In April and

May, in the Gulf of Benin and Biafra, one may be looked for every forty-eight hours, and frequently two in one day, and extremely violent. On the Ivory and Gold coasts they last till June. The period of heavy rains in the Gulf of Benin and Biafra is from August to September; but on the Ivory and Gold coasts it is from May to June.

The fogs, (commonly known as the "smokes,") which are very thick on this part of the coast of Africa, take place in July, August, and September on the Ivory and Gold coasts, also from December to February. In the Gulfs of Benin and Biafra, they are particularly found from October to February. These fogs generally commence at three o'clock in the morning, and clear away towards ten or eleven a.m.

Island of Fernando Po.—In the island of Fernando Po the climate is the same as that of the Gulf of Biafra adjacent to it. The harmattan blows there from December to February,—the most healthy season of the year. On the coasts above-mentioned and at Fernando Po, in the rainy season, the alternate land and sea breezes either fail or blow irregularly.

At the island of St. Helena, S.E. winds blow during nearly the whole year. They are only interrupted for a few days during this interval by light westerly winds, principally in June, July, and November; in which last month there are generally six days of westerly wind.

Ascension Island.—At the island of Ascension the winds are the same as at St. Helena, continuing moderate during the whole year.

Cape Palmas and the Coast of Liberia.—In the latitude of Cape Palmas, and to the South of this cape, the trade wind is W.S.W. To the North of the cape, it blows from S.W. and S.S.W., and during the fine season (from December to March) it varies from W.S.W. to W.N.W. The rainy season here lasts from May to October. The same winds prevail on the coast of Liberia.

The great rains fall principally in July and August. In April and May there are violent tornadoes, but they cease during the heavy rains and recommence in October and November. During the fine season the alternate land and sea breezes are very regular; the former are found twelve miles from the coast. The land breeze varies from N.N.W. to N.N.E., and lasts from noon to midnight. The sea breeze varies from W.S.W. to W.N.W., changing very gradually, and at the middle of the season changes to North. It changes to South at the end and beginning of the season, and shifts, according to circumstances, eastward or westward. There is often an interval of calm between the land and sea breezes. The harmattan blows in December, but only occasionally and then with no streugth. It is neither cold nor disagreeable, as on the coast to the North of Cape St. Anne.

It has been observed that on the coast of Liberia. during the rainy season, the weather is not so bad on shore as it is thirty or forty miles out at sea. Thus, in this season, at that distance, calms, rains, and baffling winds are found, and even storms and tornadoes; in both of which the wind is mostly from East, but not strong.

Coast of Sierra Leone.—On the coast of Sierra

Leone, during the fine season—from November to April, the prevailing wind is from N.N.W. to N.W. In the winter season it is S.W., changing to W.S.W. and W.N.W., blowing sometimes strong from West. The harmattan is sometimes severe in November and December; then, during the following months, it becomes moderate. It is not permanent, and blows only at intervals, varying between E.S.E. and N.E.

Tornadoes occur in May. During the great rains from June to September, they partly cease, and return in September, October, and November. In the winter season the sea breeze is generally light, changing from S.W. to W.S.W. and interrupted by N.W. winds.

During the fine season, from the Isles de Los to Cape St. Anne, alternate land and sea breezes prevail, from W.S.W. and E.N.E., *passing by the North.* The sea breeze lasts from ten or eleven o'clock in the morning till midnight. The change in the land and sea breezes takes place round by the North, after an interval of calm or only a successive change of wind from W.N.W. and N.W. to North and N.E.

In taking a general view of this coast, the prevailing winds are found to be from the westward; from W.N.W. during the fine season; and from W.S.W. and S.W. during the rainy season, or from May to November.

The following table has been compiled from numerous observations at Sierra Leone during the year.

Months.	Winds.	Remarks.
January	N.W.	Sea breeze in the afternoon; harmattan in the morning.
February	N.W.	Storms with rain.
March	N.W.	Tornadoes.
April	N.W. to S.W.	Ditto.
May	S.E. to S.W.	No tornadoes; partial storms.
June	S.E. to S.W.	Rain with S.E. winds; intense heat.
July	S.E. & S.W.	Ditto.
August	S.E. & S.W.	Ditto.
September	E. to S.W.	Tornadoes.
October	W.N.W. & S.W.	Cloudy; oppressive heat.
November	N.E. & W.N.W.	Ditto.
December	N.E. & N.W.	Ditto; thunder and lightning towards evening.

Weather of 1834 at Sierra Leone. The numbers in the columns express days.

Months.	Fine.	Cloud.	MistF.	Rain.	Months.	Fine.	Cloud.	MistF.	Rain
January.	31				July	5		3	23
February.	28				August	2			29
March	30	1			September	10			20
April	26		4		October	20		5	6
May	14		5	12	November	21		4	5
June	14		3	13	December	23		4	4

Coast and Archipelago of Bissagos.—On the coast and archipelago of Bissagos, the West winds, changing from W.N.W. to S.W., prevail during nine months of the year. They blow in the winter season (from May to October) without interruption from W.N.W. to S.W.b.W., and sometimes with violence during July and August. Tornadoes take place principally in June, and also in September and October.

The harmattan blows (and sometimes with much strength) in November, December, January, and the beginning of February. In the fine season, along the coast and in the archipelago, alternate land and sea breezes are found. Those from the land blow from N.E.

to E.N.E. and E.S.E. till eight or nine in the morning, then till eleven or twelve o'clock there is calm, which is succeeded by the sea breeze, rising from W.N.W. or W.S.W. It lasts till past sunset, and is succeeded by the land breeze, which rises towards midnight.

Coast of Senegambia.—On the coast of Senegambia, during the fine season (from September or October until May) the prevailing winds are N.E., changing to N.W. by the North. The solar breezes are settled and regular on this coast; they are mostly moderate, though occasionally strong.

The harmattan blows with violence in November, December, and January; it becomes moderate in February and March. It continues sometimes for six or nine successive days, and at other times blows only in the morning.

In the winter season violent tornadoes occur in May and June.

The great rains commence in July and last during August; and at the end of this month there is sometimes a return of tornadoes. The prevailing winds during this season are from S.W., light, and interrupted by calms; they sometimes blow strong from West. On this coast, while the fine season lasts, land and sea breezes blow alternately; the solar breezes are more regular, varying from N.N.E. to North in the morning, and from North to N.N.W. and N.W. in the evening. During the night the wind is light from East and E.N.E.

Cape Verd Islands.—Among the Cape Verd Islands from November to May, the trade wind blows from N.E. to North or N.N.W. In the three first months of the year it is generally more from the North than on the coast of Senegal. In June it is from East, and weaker. The rains begin about the end of this month. From July to October there are tornadoes and rain. During the rainy season, from June to October, South winds are found, changing to S.E. and S.W., stormy, sometimes with fog. After the 15th of August it is not prudent to remain in the bays of these islands, which are exposed to S.W. and S.E. winds.

Coast of Senegal.—On the coast of Senegal, and between Cape Blanco and Cape Verd, winds from East to N.E. prevail from October to May, including eight months of the year. The winter season lasts from June till October, when tornadoes, and light winds from S.W. to W.S.W. occur.

At some distance from the coast, in the fine season, North winds are often found blowing towards the shore, while, at the same time, further out at sea, the wind is from N.E. This coast is equally subject to the solar breezes, varying from N.E. to N.N.W. The breezes from N.N.W. prevail in the afternoon, those more easterly in the night and towards morning.

Canary Isles.—In the archipelago of the Canaries, situated near the limit of the N.E. trade winds, the winds blow from N.N.W. to N.N.E. by the North, during nearly the whole year, and particularly from April to October. From this last month also till February their direction is nearly the same. These winds are however interrupted by violent S.E. and S.W. winds, which last sometimes seven or eight consecutive days in December to January, accompanied by much rain.

The roadsteads of the Canaries are dangerous during these winds, and they ought not to be visited at such periods. In the Grand Canary, the bay of Palmas is the only one which may be frequented without danger in December and January, because a ship can get under sail there with any wind.

Madeira.—At Madeira the N.E. trade wind becomes settled about the middle of April, and continues so till the end of September. In October the periodical rains may be expected, which commonly last for fifteen days. They frequently begin with a strong S.E. wind, which changes to S.W., and continue to veer round to N.W., when the weather clears up. The roadstead of Funchal is very dangerous with these winds.

In November and December fine weather is found there, with the N.E. wind, which is yet irregular. January and February are the two months in which strong S.W. and South winds occur; but N.E. winds often blow during these months. In March the prevalent winds are generally from N.W., and sometimes very strong. During this month a great deal of snow falls on the mountains of Madeira.

In April the weather is often bad until the middle of the month, and it sometimes blows very hard; but it is mostly fine in the beginning of this month. In May, June, and July, the nights are clear and the days cloudy. Regular land and sea breezes then prevail.

During August and part of September the harmattan, called by the inhabitants the East wind, sets in; it blows sometimes from the East, during six or nine days following, as it does on the coast of Morocco opposite Madeira.

There is not a gale of wind in this island from the middle of April until the end of September: but they may be expected in November and December, commencing either to the eastward or westward of South, gradually drawing to West, and terminating at N.W.

Coast of Morocco.—On the coast of Morocco the prevailing winds during the fine season are fresh from N.E. to N.W. by the North. In winter S.W. and S.S.W. winds prevail, veering sometimes to W.S.W. and blowing hard. The wind shifts from S.E. to S.W. and W.S.W. rapidly in the winter, bringing bad weather; but when it is W.N.W. or N.W. the weather clears up.

Coast of Portugal.—On the coast of Portugal, and in general from Cape Finisterre to Cape St. Vincent, during ten months of the year northerly winds prevail. varying from N.E. to N.N.W. They blow fresh with fine weather, especially during summer. If there should be a gale during winter, it comes most frequently from South or S.W., sometimes from W.S.W., and blows very hard.

From Cape St. Vincent to the Canary Isles, the prevailing winds are from N.E. to N.W.

Bay of Biscay.—In the bay of Biscay the wind is most variable; but it has been observed that in the winter months it varies from S.W. to N.W. by the West, the last being the most frequent. From May to September, sometimes also in December and January, winds from E.N.E., East, and E.S.E. are found. During the two last-mentioned months. these winds are fresh and lasting; those from N.E. freshening up with rain, and if there is a gale of wind it will come from East or S.E., and may be expected to be severe.

On the coasts of Brittany S.W. winds prevail; weak in summer, but violent in winter, and changing from West to N.W., from whence heavy storms and gales of wind may be expected.

At the entrance of the Channel, and on the West coast of France, the prevailing winds are generally from S.W., varying to West, W.N.W., and N.W. They last very long, blowing for seven or eight months, and freshen into violent gales, especially in winter. The wind from W.S.W. and S.W. is generally accompanied by rain or fog, while from N.W. it is stormy, but attended frequently with a clear sky. Should the N.W. wind be moderate, it is generally attended with fine weather, interrupted, however, in winter, by storms of considerable violence, attended with hail and thunder. On the West coast of France these are commonly called "sea storms." These winds may be relied on more than any others. They sometimes originate in northern America, and traverse the whole Atlantic Ocean.

In summer S.W. winds prevail, alternately moderate and fresh with foul weather. However, in this season, the sky is generally clear with a S.W. wind,

In this season, if the wind is S.W. and the weather fine, if it veers to N.W. it generally strengthens, but the weather will still continue fine.

Strong S.W. and N.W. Breezes in the Bay of Biscay.—In the English Channel or Bay of Biscay, when the wind comes in squalls from S.W., whether in summer or winter, if it be attended by rain, increasing in quantity, and the squalls become heavier and more frequent, with a slight tendency to vary, a change of wind may be expected. Generally the change is from S.W. to West, rapidly, and sometimes to N.W. in a squall, and blowing harder than before. This state of weather may become of serious consequence to ships working to windward and on the port tack, and even to those running free or with the wind abaft, if unprepared and sail be not reduced in time.

Sometimes the change from S.W. to N.W. is preceded by a short calm, which must never be trusted. It has also been observed in the bay of Biscay that when a breeze springs up from a point opposite to the sun, it does not last, and indicates merely a slight derangement of the atmosphere.

Winds from North and South.—Winds from North and South are not very frequent; they prevail now and then, but not to a great extent nor for a long interval, although they sometimes freshen up into strong breezes, and even gales. Those from South will draw to S.E. or S.W.; and those from North will become N.E. or N.W.

Entrance of the Channel.—At the entrance of the Channel, although the wind is very changeable, it is found that westerly winds are most prevalent in September, October, and November; and easterly winds in December, January, and February.

British Isles.—On the shores of the British Isles, the prevailing winds are nearly the same as those in the bay of Biscay. It is, however, observed that in Scotland North winds are very frequent, and that East winds blow principally from March to June. In England and Ireland South and S.W. winds prevail; and it is observed that on the coast of Cornwall westerly winds blow during nearly nine months of the year.

Coast of Holland.—On the coast of Holland westerly winds prevail, attended by rain and fog. Winds from S.E., South, and N.W. seldom blow, but from S.W. they do frequently; and easterly winds, which occur in every month of the year, are found principally in the four winter months, producing dry cold weather.

Coast of Norway.—On the western coast of Norway, the prevailing winds are from S.W. to South, frequently attended by rain.

Greenland.—Greenland cannot boast a periodic wind. From May to July the weather is fine, the wind changeable, but most frequently very strong from S.S.W. Even until September the winds are variable, but rain is by no means frequent. Storms seldom occur, and when they do are of short duration. The most violent squalls come from South. The coldest winds are from N.E., and this may form the sum total of our remarks on Greenland.

Arctic Region in North America.—In the following table we have resumed the observations on winds collected by Captain Parry in his voyage to the arctic regions to discover a passage from the Atlantic to the Pacific Ocean. These observations are doubly interesting since they are continued without interruption from July, 1819, to September, 1820; consequently showing the prevailing winds of these frozen regions during more than a year. This table is only a summary of those

Month.	North.	N.N.W.	N.W.	W.N.W.	West.	W.S.W.	S.W.	S.S.W.	South.	S.S.E.	S.E.	E.S.E.	East.	E.N.E.	N.E.	N.N.E.	Calm.	Variable.	General Remarks.
1819.																			
July .	3	5½	3	..	4½	..	2½	..	1	2½	3	..	2	..	..	1½	2½	..	E. and S., fresh, with rain; the other breezes light; much mist and fog; snow with N. wind.
Aug. .	4	..	5	..	1½	..	6	..	2	½	1½	..	4	..	2½	1½	1	1½	Strong E. winds and fog; N., moderate, cloudy; S.W., rain; N.N.W., fresh, cloudy, misty.
Sept. .	7	4	4	..	3½	..	5½	..	..	..	..	..	..	..	1½	4½	..	..	Strong W. in gusts; N.N.E., fresh, misty; snow with strong N. wind; other breezes moderate.
Oct. ..	10½	7½	1	..	5½	..	3	..	..	1	..	..	1	..	1	..	..	½	Strong N., weather clear, fine; N.W., fresh, cloudy; W., fr., snow, fog; S.W., st., misty; E., st., snow.
Nov. .	16	6	1	..	1½	..	1½	..	..	1	..	..	..	..	..	..	1	2	N., moderate, fine, snow, hurricanes; S.W., fine; other breezes light or moderate.
Dec. .	5½	3	5	..	2	1	½	..	3½	..	1	3½	5	..	..	..	..	1	E., fresh, weather cloudy; mist and fog with S.
1820.																			
Jan. .	11½	7½	½	..	4½	..	..	..	..	1	1½	1	2½	..	..	1	½	..	S.S.E., st., fine weather; N. and N.N.W., st. or fr., fine; much fog during this month.
Feb. .	9	9	..	..	3	..	..	..	1	..	1	..	1	..	..	..	2	3	N.N.W., strong; N., fresh, much fog.
March	17½	4	2½	..	2	..	..	..	1	½	..	1	..	..	..	..	..	2	N., fr., weather clear, some snow; S., clear; W. fr.
April .	9½	4	3	..	..	..	..	..	..	½	1	2	2½	..	..	1½	1	5	N.N.W., fr.; N., light, snow, fine; E., fr., snow.
May..	11	3	5	½	..	..	..	..	4	½	..	1	1	..	1	..	3	1	N. and N.N.W., strong, alternate clear and cloudy weather; squalls from N.W.
June .	7	2½	2	..	4	..	..	2½	2	1	3½	..	1½	..	..	..	..	5	N., st., fine, cloudy, rain, fog, snow for two days.
Total	111½	50	32	½	32	1	19	2½	14½	8½	12½	8½	19½	..	6	10	11	21	
July .	9	1	1	3	1½	..	1½	1	4	2	1	..	..	..	..	3	..	3	N., fr., cloudy; S., fr., rain; S.S.W., st., freq. fog.
Aug. .	1½	1	3	3	8½	3	..	1	..	..	..	2	1½	1	..	..	1½	5	W.N.W., N.N.W., E.S.E., fr., snow, fog, often cldy.
Sept. .	1	2½	4½	1	1½	2	1	..	4	1	5	..	1	..	2½	1½	..	..	S.W., strong, seven days snow, mist, fog; varying breezes, fresh, and often in gusts.
Total	11½	4½	8½	7	11½	5	2½	2	8	3	6	2	2½	1	2½	4½	1½	8	

The figures indicate the number of days during which the wind has blown from the quarter stated in the first horizontal column.

published in Captain Parry's voyage, who from September, 1819, till August, 1820, remained between the parallels of 74° and 75° North latitude.

Summary of observations made on board the *Hecla* during an interval of twelve months, in which period the vessel was in the latitudes of 74° and 75° N.

Months.	Thermometer.			Barometer.		
	Max.	Min.	Mean.	Max.	Min.	Mean.
1819.	°	°	°			
September ..	+37	− 1	+22·54	30·42	29·36	29·90
October	+17·5	−28	− 3·46	30 32	29·10	29·81
November ..	+ 6	−47	−20 60	30·32	29·63	29·94
December ..	+ 6	−43	−21 79	30 75	29·10	29·86
1820.						
January	− 2	−47	−30·09	30·77	29 59	30 07
February ...	−17	−50	−32·19	30·15	29·32	29 76
March	+ 6	−40	−18·10	30·26	29·00	29 80
April	+32	−32	− 8·37	30·86	29·40	29·97
May	+47	− 4	+16 60	30·48	29·25	30·10
June	+51	+28	+36 24	30·13	29·50	29 82
July	+60	+32	+42·41	31 01	29·13	29 66
August . ..	+45	+22	+32·68	30·03	29·46	29·73

Remarks.—The thermometer when placed on shore or on the ice at a distance from the ship invariably stood from 3° to 4° or 5°, and even on some occasions 7°, lower than on board. The mean temperature for the year may therefore be fairly considered as —2°. The lowest temperature registered on the ice was —55°; it did not rise above —54° for seventeen hours on the 14th and 15th of February, 1820.

The two preceding tables conclude the observations on the Arctic region of North America, and, in order to render them as complete as possible, we further give the table of observations also made by Captain Parry in his third voyage to discover the North-West passage. These observations embrace a period of sixteen months—from June, 1824, to September, 1825.

Observations made at Port Bowen in lat. 73° 49′ N. and long. 87° 25′ W.

Month	North.	N.E.	East.	S.E.	South.	S.W.	West.	N.W.	Var.	Calms	Remarks on the Weather.
1824–5											
June	½	2	11	1	8½	3	..	2	2	..	Much fog and rain; fine with E. wind.
July	9½	5	½	5	2½	1½	1½	5½	..	..	Wind variable and weak; foggy.
Aug.	3½	1½	3	7	3½	2½	1½	7	1½	..	Remarkable for rain and snow.
Sept.	1½	1½	4½	9	..	2	9	1½	..	1	Breezes fresh and gusty.
Oct.	4	2½	10½	6½	1	..	½	5	1	..	Fresh E. breezes; snow and fog.
Nov.	2	..	8½	8	..	2½	1½	5½	2	..	Wind by gusts; clear with N.W. wind.
Dec.	4	1	15	4½	..	1	1½	2	2	..	Fine with E. wd.; clear, a little snow.
Jan.	3	2¼	18	1½	..	..	..	3	3	..	Very fine month; few storms.
Feb.	3	2	18½	..	..	..	1	2½	..	1	Ditto.
Mar.		½	18	..	..	2	2	7½	1	..	Fine; some storms.
April	2½	..	18	2	..	..	4	2½	1	.	Fine; a few storms; a little snow during five days.
May	3	3	9½	1	3	1	2	7½	1	..	Strong breezes from S.E.; dull weather, squalls, snow.
June	1½	1	12½	2½	2	4½	2½	3½	..	..	Stormy; wind variable; misty.
July	4	1	..	5	2½	3	12	4	..	..	Weather generally cloudy.
Aug.	9	8½	1	..	1½	2½	1½	7	..	..	Cloudy, mist, rain, light breezes.
Sept.	1½	3	3½	1	5½	3	4½	6	..	..	Cloudy, rain, fog, light breezes.
Total	52½	35	158	54	30	28	45	74	14½	2	

This may conclude the Arctic observations with the remark that in the Arctic region the winds are variable and irregular; while they are generally moderate in all seasons of the year.

Hudson Bay.—In Hudson Bay it is observed that from October to May the prevailing winds are from North to N.W., and from June to October from S.E. to East. The northerly winds are very strong, and in spring and autumn squally tempestuous weather is most common.

Canada.—According to some writers, we find in Canada winds blowing regularly from North during five winter months. According to others, N.E. and S.W. winds prevail alternately; the former at the end of autumn and during winter, the latter during the remainder of the year. From December to April, the weather is generally serene. The occasional N.W. winds which blow at this period are colder than those from N.E., and are common while the ice lasts. They are only met at sea in these regions about the month of March. They increase in June, and afterwards gradually diminish. The following table contains a summary of the winds observed during the year 1834 in Lower Canada.

Months.	Winds.	Remarks on the Weather.
January	W.N.W.	Weather generally fine.
February	W. and E.N.E.	Much snow.
March	W. and E.	Snow and rain.
April	Variable.	Generally fine.
May	Ditto.	Ditto.
June	Ditto.	Ditto.
July	Ditto.	Ditto.
August	Ditto.	Ditto.
September	Ditto.	Rain and cloudy weather.
October	E.N.E.	Snow and rain.
November	S.S.E.	Snow.
December	W.N.W.	Variable.

Newfoundland.—On the East and South coast of the island of Newfoundland the winds most generally found are from South from May till October. They are, however, very changeable, and generally moderate during this period. Nevertheless, there are occasional squalls from S.E., with rain and fog; which latter is especially prevalent in July and August. The N.W. winds which occasionally blow are dry and cold, and generally attended with a clear sky. In October these winds become violent. S.W. winds are also found here, but very variable in strength, at all times of the year.

Nova Scotia.—The following is a summary of observations of the winds at Halifax on the East coast of Nova Scotia.

Months.	Winds.	Remarks on the Weather.
January	N., S., and W.	Clear, rain, snow.
February	N.W. and var.	Clear, rain, cloudy.
March	N.W. and S.W.	Ditto.
April	West.	Ditto.
May	N. and W.	Clear and rain at intervals.
June	W., N., and N.W.	Ditto.
July	W., N., and S., var.	Clear and misty.
August	W. and S., var.	Clear, cloudy, rain, mist.
September	N.W. and S., var.	Ditto.
October	S.W., N., and N.W.	Clear.
November	W. and S.W.	Clear, rain, mist.
December	N.W. and N.E.	Clear and rain.

New Brunswick.—In New Brunswick the following observations have been made at Fredericton, the capital of this colony, by Sir James MacGregor.

Months.	East.	South.	West.	North.	Var.	Fine.	Rain.	Mist.	Snow.
January	4	..	7	6	14	24	2	1	4
February	2	4	4	2	16	23	1	..	4
March	23	2	5	..	1	22	2	2	5
April	12	4	11	..	3	22	7	..	1
May	20	1	7	..	3	18	8	5	..
June	19	1	10	..	..	15	6	9	..
July	20	..	7	2	2	18	3	10	..
August	17	..	9	4	1	23	3	5	..
September	17	..	10	2	1	17	5	8	..
October	14	..	8	..	9	22	7	2	..
November	11	5	..	14	..	15	8	3	4
December	..	..	9	14	8	26	..	2	3
Total	159	17	87	44	58	245	52	47	21

East Coast of North America.—On the East coast of North America the wind is neither constant nor uniform. The limits of the trade wind extend more northerly on this coast than on the opposite one of Africa, and reach the parallel of the Bermudas, and sometimes even as far as 32° North latitude. The wind becomes more easterly as it nears the coast. In this part of America, which comprises the United States, the most frequent winds are from N.W. to S.E. In winter they are generally from N.W., and are most frequently dry. Winds from East, E.S.E. and S.E., produce rain, and the latter often very severe.

Cape Hatteras is celebrated for the constant bad weather it enjoys during the greatest part of the year. On this coast, says Franklin, hurricanes from N.E. are found; they first visit the S.W. part of the United States, in Georgia, and from thence extend successively ever the country in their progress to the North, sometimes reaching Newfoundland. These violent winds last sometimes for two or three days, accompanied by rain and dense clouds.

Gulf of Mexico.—In the interior of the gulf of Mexico the trade winds are generally found, but only near the coast in proportion as the local winds found there diminish their force.

The Northers.—From May to September, in the gulf of Mexico, during June, July and August, is the season of the Northers: these winds are extremely violent, and sometimes accompanied with rain. The most violent are called *Hueso Colorados;* the more moderate *Chocolateros;* and they are found as far as the Bahama Channel. The Northers of the gulf of Mexico are announced by a heavy swell getting up in the bay of Campeche, by considerable humidity in the atmosphere, and by a dark cloud seen in the N.W. in the morning and evening, keeping 9° or 10° above the horizon, for two or three days sometimes, before the Norther arrives; lightning in the N.W. and N.E., and gossamer floating in the air and hanging in the rigging, called *hilos de la vierga;* all these, as well as the phosphorescence of the sea, are indications of the approaching Norther. The wind commences first with a light air from South,—and then makes the round of the compass by West, and when it arrives at N.N.W., blows with all its violence. These breezes, which are very dangerous in the gulf of Mexico, generally last two or three days. When the wind is N.W., if the black cloud abovementioned begins to disappear, the gale will last but a short time, and the wind, which is then much less, hauls to the East, and if it becomes N.E. the weather moderates. These breezes are often attended with much rain, and heavy cloudy weather, and they necessarily produce a very heavy sea.

Dry Season.—The Northers of the gulf of Mexico prevail from the middle of March till September, which is the dry season. These winds in September and October are stronger than at other times, and if they are not found at this period, the trade winds are interrupted by storms and rain. In November they are quite settled, with considerable strength, and continue during December, January and February. In March and even April, when they blow they are light, and are then stronger at daybreak than in the preceding months; they also *veer to the N.W.*

Damp Season, or Season of the Trade Winds.—The damp season on the coast of the gulf of Mexico, prevails from March to September. From the end of March and during April the trade winds from E.S.E., interrupted from time to time by North winds, are sometimes attended with a clear sky, at other times with a sky overcast; they veer towards S.E. and last all night. From July to October, the period when the North winds prevail, violent storms, accompanied by thunder, lightning, and heavy rains, are frequently experienced. Those from East are the most severe, but of the least duration.*

Hurricane Season.—The period of hurricanes in the gulf of Mexico, like the Antilles, is principally from August to October;† and the rainy season, called the winter in these regions, like the opposite on the coast of Africa, commences when the sun reaches the zenith of the place, passing to the North; and terminates when it again reaches the zenith of the same place, passing to the South.

Bahama Channel.—In the Bahama Channel the trade wind from N.E. in winter is interrupted by North winds, and in summer by calms. In winter, that is to say, from November to April, we find winds changing from East to South, and from South to West. In December and January there are frequently North winds, changing to N.W., blowing violently for seven or eight consecutive days.

In summer, from May to September, the prevailing

* Bernardo de Orta: *Derrotero de las Antillas.*

† We do not allude now to the hurricanes of the Atlantic Ocean. In the Indian Ocean we have collected all the facts relative to these terrible phenomena, and pointed out the laws observed by authors in their progress. See *Storm Compass: by A. B. Becher, Captain, R.N.*, published by Potter, 31, Poultry.

winds of the Bahama Channel are from S.E. to S.W. by the South. During March and April southerly winds are frequent.

East Coast of Florida.—From the parallel of 28° North latitude to the Florida Cays, the trade wind blows till noon, and shortly after is followed by the sea breeze. This takes place regularly during summer: in winter, principally from November to March, the winds blow from South to West, and bring a heavy sea.

West Coast of Florida.—On the West coast of Florida, even as far as 28° North latitude, alternate land and sea breezes prevail.

Pensacola.—At Pensacola, chiefly from April to July, in the morning the wind is from North to East, or from East to South, followed in the afternoon by the wind from S.W. These S.W. winds (from sea) are termed *Virazones;* they blow in squalls in August, September, and October, a period when hurricanes are also experienced. From November to March northerly winds prevail; they begin at S.E. and South, with much rain; then veer to S.W. and West, at which point they remain some time, blowing very strong till they shift to N.W. and North; the weather then becomes fine.

Coasts from the Mississipi to the Bay of San Bernardo.—From the mouth of the Mississipi to the bay of San Bernardo, land breezes are found at daybreak from April to August. A short time after daybreak the wind veers to East and S.E., and blows from S.W. in the afternoon. In July, August, and September, squalls are frequent, with rain; there are, besides, southerly winds, changing from South to S.W. in squalls, which last several successive days. The worst months for navigating this coast, are those of August, September, October, and November, because the winds are severe, and blow dead on shore, without permitting a vessel to carry as much sail as will take her out to sea. In February, March, and April, there is much fog at the entrance of the Mississipi. From December to March there are frequently strong northerly winds; and if these winds veer to East or South of East, the weather becomes dark, cloudy, or foggy.

Coast from the Bay of San Bernardo to Tampico.—Between the bay of San Bernardo and Tampico winds from S.E. continue from April till August; during the other months of the year strong winds from East and E.S.E. are found on that coast, lasting two or three days before blowing from North. During the fine season, the land breezes prevail regularly from eleven or twelve at night to nine or ten in the morning.

Tampico to Vera Cruz.—On the coast from Tampico to Vera Cruz, from April to July, the winds during the day blow from East, changing to E,S.E.; during the night they veer to South and S.W.; that is to say, they blow off the land. If the land breeze, on the contrary, shifts to N.W., accompanied by a little rain, the next day the wind generally comes from North of N.N.E. or N.E., especially in August and September. The land winds are termed *Vientos de cabeza,* (head winds,) or *Vendavales.* These winds are generally light: they do not reach further than twenty or thirty leagues from the coast, at which distance they blow from East or E.S.E.

Vera Cruz.—At Vera Cruz the winter season commences towards the middle of May, and terminates towards the end of July, when there are frequent interruptions in the trade winds, much fog, and stormy weather. From the end of July to the middle of October, the period when northerly winds prevail, storms are heavy.

The most violent winds are from East, but do not last long. Northerly winds prevail from the middle of October till March, but they generally fall at sunset, blowing strongest from nine o'clock in the morning till three in the afternoon. This is not the case if the wind only gets up in the afternoon or evening, in which case it continues to blow during the night, gradually increasing in force. At night, and after midnight, the wind changes, and shifts to N.W., blowing from the land. In this case if towards morning it veers to S.W., the North wind will not last, and the sea breeze will follow at the usual time, about nine or ten in the morning; but if this does not take place towards sunrise, or at latest at the beginning of tide, the North wind will again blow with as much force as on the preceding day; it is then called the North tide wind. The North wind often terminates by shifting to East, a guarantee of fine weather. If in the afternoon it veers to N.E., the sky will be cloudy the next morning; when the land breeze has come from South to West during the night, a sea breeze may be expected in the evening.

The weather then continues fine during five or six days; the longest period of fair weather with northerly winds. In case of the winds backing from N.E. to

N.N.E. and North, the weather is uncertain.* An attention to these peculiarities of the winds is important as affecting the approaches to the coast of Vera Cruz.

Coast of Yucatan from Vera Cruz to Point Arenas.—On the coast of Yucatan, from Vera Cruz to Arenas, we find, during the dry season, alternate land and sea breezes. The sea breeze from North, the land breeze from South, from seven or eight o'clock in the evening till eight or nine in the morning. The dry season lasts from September till April or May. The rainy season follows, and continues till September; it is announced by tornadoes and violent storms, which become more frequent in May and June. The great rains fall in July and August, being then continuous and very heavy. In this season there are sometimes strong winds from E.S.E., lasting three or four days.

Winds from North to N.E. begin in October; they are very strong in December and January, and gradually fail towards March; in general they are fresh and dry, and stronger than common breezes.

Coast from Point Arenas to Cape Catoche.—On that part of the coast comprised between Point Arenas and Cape Catoche, the seasons are nearly the same as the foregoing, only the general winds are from N.E., interrupted by strong North winds. In April tornadoes occur from N.E. to S.E. This squally weather lasts till September, during which sea breezes set in from N.N.W. to N.W. These breezes rise at eleven o'clock in the morning and during the night veer to East and E.S.E.,

* Bernardo de Orta: *Derrotero de las Antillas.*

and afterwards to S.E.; which last may therefore be regarded as land breezes.

It has been observed here, that the stronger the wind is from N.N.W. to N.W. the more violent are the tornadoes. On this part of the coast it has also been remarked that the rainy season is shorter than on the neighbouring coast westward.

Antilles.—The N.E. trade wind prevails particularly over that portion of sea called the Caribbean Sea. When approaching the shores of these isles, however, disturbances are found in these winds. Thus: on the shores of the Great Antilles, Cuba, Jamaica, St. Domingo, and Porto Rico, the sea breeze blows regularly during the day, and the land breeze during the night. The land breezes are fresher than those observed near the coast, and are favourable for making passages from West to East in this sea.

In the Lesser Antilles the land breeze is not met with, or if it should be it is at so short a distance from the shore as to be useless in navigation. In these islands two seasons are observed,—the dry and the rainy. Their periods vary in the different islands; but it may be stated generally, that the first lasts from October to June, and the second from June to October. During the dry season the N.E. trade wind blows regularly and fresh, with a clear sky. From June to October (the winter season) tornadoes and severe hurricanes are experienced, between the 15th of July and the 15th of October. In the Antilles the hurricanes blow from the West. This is a necessary consequence of the focus of the hurricane passing to the northward of them, as it usually does.

Hurricanes rarely penetrate into the gulf of Mexico; some, however, have crossed the gulf, and continued onward beyond Vera Cruz. Amongst others, those of the 18th of August, 1810, and June 23rd, 1831.

We quote the following statement of the course of the hurricane in the Atlantic from a little pocket treatise on Hurricanes. "In the Atlantic Ocean, it is shown by Redfield that their average place of commencement is in the latitude of 15° N., and longitude 55° W., or about N.E. from the island of Trinidad. From thence they pursue a W.N.W. course, until arriving near the coast of Florida they follow the course of the Gulf Stream to the N.E., sweeping past the coast of the United States, and continuing far beyond the eastern limits of Newfoundland. Some, originating South of those, have maintained their western course beyond the gulf of Mexico; and others, again, North of them, have assumed their N.E. course, passing between Bermuda and the American coast. But the North Atlantic hurricane mostly commences N.E. of Trinidad, within the parallels of 10° and 20° N. and between 50° and 60° West longitude."*

Calms and Storms.—Under the wind off the high land which forms the greater part of the Antilles, calm is experienced, interrupted by violent and very dangerous squalls, coming down from the declivities of the mountains; and it is only at two or three leagues off at sea that the regular breeze is again found. These tricks of

* *Storm Compass; or, Seaman's Hurricane Companion, by A. B. Becher, Captain, R.N.*, p. 5. Potter, 31, Poultry, London.

the wind announce their approach by a shrill whistling. and sometimes by an agitation of the surface of the sea. They must not be trusted in sailing by the wind from the islands, and great care is necessary in looking to the sails. Vessels have been dismasted by these breezes, and many have been capsized even at the entrance of the bay where they had intended to anchor. During winter strong tide rips are found in most of the bays of the Antilles, generally after calms or light airs.

Cuba.—At Cuba the rainy season is from June to September, and the N.E. trade wind blows over the whole island from March to October. During the other months it takes frequently a northerly direction, changing to N.W.; it is then sometimes very strong. In the fine season, the sea breeze is regular on the North coast; it commences towards eleven o'clock or at noon, and towards evening gives place to the land breeze. It is found, however, that the trade wind prevails on this coast of the island, and that winds from South to East frequently prevail in the morning, and shift from E.N.E. to N.E. towards evening. At Havana the sea breeze generally sets in about ten o'clock in the morning.

On the South coast there are alternate land and sea breezes, the land wind commencing shortly after sunset.

The Bahama Isles.—The following observations of one year were made at Nassau, in the isle of Providence:—

Months.	Winds.	Remarks on the Weather.
January	S., N.E., N.N.E.	Strong breezes and cloudy.
February	N.E., S.E., N.E.	Moderate and variable.
March	N.E., S.E., N.E., N.	Clear, but breezy.
April	E., N.E., S.E., N.W.	Clear, a little rain.
May	Variable.	Moderate, showery.
June	Ditto.	Clear and dry.
July	S.E., E., S., N.E.	Light and clear.
August	N.E., E., N.W., S.	Squalls, with rain.
September	Ditto.	Clear, rain, and fog.
October	E., N.E., N.W.	Light, rain, and squalls.
November	S., S.W., W., N.W.	Moderate and squally.
December	S., S.W., N.W.	Variable, light, clear.

Jamaica.—In the island of Jamaica the alternate land and sea breezes are well established during the fine season; the former extend four leagues from the coast and cease towards four in the morning. The following are observations made on the winds of this island:—

Months.	Winds.	Remarks on the Weather.
January	N. and S.E.	Fine; small rain; strong winds from N.
February	Ditto.	Fine and dry; strong sea breezes.
March	Ditto.	Ditto.
April	Ditto.	Very dry; breezes moderate.
May	Ditto.	Fine; some showers.
June	Ditto	Generally fine: heavy rain.
July	Ditto	Much rain; fine by intervals.
August	S.S.W.	Some heavy rains.
September	S.S.W. and S.E.	Fine mornings; rain in the afternoon.
October	Ditto.	Heavy rain by intervals; generally fine.
November	Ditto.	Ditto.
December	Ditto.	A little rain; generally fine.

On all the coast of this island alternate land and sea breezes are found, and their directions vary according to that of the coasts. The sea breeze commences about eight or nine a.m., increases till noon,—sometimes till four p.m., and afterwards diminishes, to give place to the land breeze, which ceases towards four or six o'clock in the morning.

Porto-Rico.—At Porto-Rico rain falls from June to August. The trade winds blow from N.E. The sea breeze commences at eight in the morning and lasts till four in the afternoon, when it is followed by the land breeze.

St. Domingo.—The winds in the island of Hayti or St. Domingo vary on different parts of its shores. The winter season lasts from the end of April till November. During this season squalls of wind and storms are frequent; and strong winds from S.E. are found in the bay of Gonaives and in the channel of St. Mark. In November, December, January, and February northerly winds, changing to N.W., blow violently, principally on the North coast of the island. On the South coast, in June, July, and August, there are frequent storms, with the wind, from South, blowing violently. On the shores of this island, the land wind, which is generally very light when it does blow, is not to be depended on.

Lesser Antilles.—The following observations made at Trinidad and Dominica will give a general idea of the winds met with in the Lesser Antilles. The first table is for the Isle of Trinidad:—

Months	Winds.	Remarks on the Weather.
January	E., E.N.E., E.S.E.	Cloudy, rain.
February	E. and E.N.E.	Cloudy, heavy dew.
March	Ditto.	Fine, dry.
April	E.N.E.	Fresh breezes.
May	S.E. and E.N.E.	Strong winds, thunder.
June	E.N.E. and E.S.E.	Rainy, breezes.
July	E.N.E.	Tempests, rain, and storms.
August	E.S.E.	Heavy gusts of wind, rain, and storms.
September	Ditto.	Heavy rain, storms.
October	Ditto.	Strong breezes.
November	E. and E.N.E.	Fine and warm occasionally.
December	Ditto.	Cold.

The following table is for the Island of Dominica:—

Months.	Winds.	Remarks on the Weather.
January	E.N.E. and N.	Cloudy.
February	E.N.E. and S.E.	Cold.
March	N.E. to S.E.	Fine, sometimes cloudy.
April	E.N.E., S.E., S.	Fine, moderate breezes.
May	N.E. to S.E. and E.	Calm, weather clear.
June	S.E., E., to N.E.	Calm, fog, and rain.
July	Ditto.	Calm, nights cold.
August	S.E. and N.E.	Calm, nights cold, storms and gusts.
September	S. and S.E.	Generally fine, rain at intervals.
October	N.E. to S.E.	Cloudy and fine alternately.
November	Ditto.	Fine and dry.
December	Ditto.	Fine, dry, and cold.

East Coast of Yucatan, Vera Paz, Honduras, and Mosquito.—On the eastern coast of Yucatan, that of Vera Paz, Honduras, and Mosquito, which form the western coast of the Carribbean Sea, the N.E. trade wind prevails in February, March, April, and May, but is sometimes interrupted (principally during the two first months) by northerly winds. In June, July, and August the winds on these coasts vary from East to West by the South, attended by squalls and calms. In October, November, December, and January the winds are from South to North, changing by the West, with squalls from W.S.W. to W.N.W., shifting to the North.

On the coast between Cape Gracias a Dios and Cape La Vela the winds are very changeable. From March to November they blow from N.E. to East, but are often interrupted by tornadoes in May, June, and July. Between the months of October and March, particularly in December and January, the wind occasionally comes from West; it is not strong, but sometimes lasts during seven or eight days, and is then followed by N.E. winds.

While the westerly wind is strong on this coast, lasting for some days, the trade wind from East nevertheless is blowing out at sea as at other times. It is met with at a distance of eight or ten leagues from Cape La Vela at the same time that the westerly winds are blowing on the coast near this cape.

Porto Bello.—At Porto Bello, and between this point and Cartagena, the wind is N.E. from the 15th of November till the 15th of May. At the end of May it veers to S.W. and W.S.W., and reaches as far as lat. 12° N. These winds from S.W. and W.S.W., which are sometimes very violent, bring rain. At twenty leagues seaward from Porto Bello the wind blows from South in the interior of the gulf, veering to N.E.; the South winds generally extend eight or nine leagues from the coast. The winds from S.W. and W.S.W. are termed *vendavales*.

Cartagena.—At Cartagena during the fine season, from the middle of December till the end of April, the wind is generally from N.E. From May till November (the winter season) rainy and stormy weather prevails. During the fine season N.E. winds become settled towards the 15th of November. In the rainy season S.W. and W.S.W. winds extend as far as lat. 12° N.; beyond which the wind takes a N.E. direction. In November and December there are strong breezes, with much rain. During the winter season tornadoes are frequent on the coast.

Coasts of Caracas and Cumana.—The trade winds

take their usual course on the coasts of Caracas and Cumana as far as Cape La Vela; but from this cape to the point of St. Blaize their direction varies from N.E. to N.N.E. During the months of March, April, May, and June they are more regular, blowing with great violence from E.N.E. These strong breezes extend from the middle of the channel to within two or three leagues of the land, and diminish in force as they near it. On these two coasts, and even as far as the gulf of Nicaragua, the rainy westerly winds to which we have already alluded, called vendavales, are found from July to December, and sometimes till January.

Coast of Guiana.—On the coast of Guiana the trade wind only is found. From January to March it is from N.N.E. and E.N.E. In April, May, and June there are variable winds and calms. Afterwards the trade veers to E.S.E. and S.E., blowing from S.E. principally between June and December. The dry season is from January to June, and the rainy season is attended by continual storms.

The following table shows the state of the winds and weather at Demerara:—

Months.	Winds.	Remarks on the Weather.
January	East.	Cold, fresh breezes.
February	N.E.	Thick clouds with tempests, stormy.
March	E.N.E.	Clouds, very heavy showers.
April	East.	Heat, no rain.
May	N.E.	Thick clouds, frequent lightning, rain.
June	S. and Variable.	Hot, and rain at intervals.
July	East and South.	Hot and very heavy.
August	South.	Hot, rain at intervals.
September	South and East.	Heat, thunder and lightning.
October	Variable.	Light breezes, showers.
November	North and East.	Ditto.
December	N.N.E.	Heavy rain, breezes cold.

On the coasts of French Guiana the winter season begins in November and ends in July; the dry season lasts from July to November. Winds from E.N.E. blow during the first period; those from E.S.E. during the second. The rainy season is from December to February, and sometimes till March; it even begins sometimes sooner,—about the 15th November. In March and April there is an interval of three weeks or a month during which time the rains cease. This period in French Guiana is termed the March summer. The rain begins again towards the middle of April, and terminates in the middle of July. The interval between the rains is called the *poussiaière*. From November to March the winds are from N.N.E. to N.E.; during March and April they change from East to South; from May to June they return to N.E., calms then are rare and there is no land wind. It is observed at Guiana that the dry winds are from East to South.

North Coast of Brazil.—On the North coast of Brazil, as far as Cape St. Roque, the trade winds blow from N.E. to S.E. From July to December those from S.E. to East prevail; from December to July those from N.E. to East. The month of June is the time when these periodical winds change, and which is the cause of the calms found near the land in that month, sometimes interrupted by squalls of wind and rain. The heaviest squalls are with the wind from East to N.E. At a short distance from this coast a land wind is often found during the night and morning, varying from S.S.E. to South, and ceasing towards eight or nine a.m.

East Coast of Brazil.—On the East coast of Brazil the winds are periodical. From September to March they blow from E.N.E. to N.E.; from March to September from S.S.E. to E.S.E. These winds do not extend more than forty or fifty leagues out to sea. Beyond this limit the trade wind is found, which generally blows between S.E. and East. On this limit, however, variable winds are met, between S.S.W. and S.E., with rain and storms. On the North part of this coast rain is frequent, as well as variable winds, in March and September,—the times when the change of winds takes place,—accompanied by heavy squally weather.

Bahia de Todos os Santos.—At Bahia the wind is E.N.E. from September till April. From April till August it is from South, changing from S.E. to S.S.W. In April it begins shifting to South and S.S.W. It is in greatest force during May, June, July, and August, and in these months is from S.E., varying to South and S.S.W. The seasons change in April and September. The wet season at Bahia begins in April, the same as on the coast of Brazil. The fine weather returns in September.

Rio Janeiro.—At Rio Janeiro the sea breeze, which comes from East, begins at eleven a.m., and reaches the roadstead and town towards two or three p.m., although the latter is only three leagues from the shore. The sea breeze lasts till sunset. The land breeze commences towards evening and lasts till morning, its duration and force depending on the season of the year.

Paraguay.—On the coast of Paraguay the sea breeze

sets in at nine or ten in the morning, and lasts till sunset.

River La Plata.—The winds in the River Plata and at the mouth of it follow the course of the seasons, but the form of its shores and their proximity exercise so great an influence on their force and direction that they are rarely the same as in the interior of the river. Thus sometimes a violent wind is blowing at Buenos Ayres which is not felt on the shore immediately opposite.

Almost all the pilots attribute great influence to the phases of the moon, and agree that it is difficult to foretell the weather correctly, the changes of the atmosphere being so sudden as to defy all their predictions. Storms gather and come down so rapidly that it is necessary to be always on guard against them. Instances of violent storms, called *pamperos*, are cited as coming suddenly when the weather was fine and clear, and announced only by a whirlwind: these, however, come only with a N.W. or westerly wind.

In this country the wind from S.W. is termed *pampero*. It is generally introduced by thick black clouds, which appear to roll hurriedly over each other; at other times by a large dark arch which invades the whole sky from West to East. The horizon quickly clears towards the S.W., and it is then that the pampero bursts forth with indescribable impetuosity. It is frequently accompanied by thunder, lightning, and rain; the coldness of the temperature is quite uncomfortable. The sky soon, however, becomes clear and the weather fine, and continues so during the rest of the pampero.

When the wind ceases it almost always veers to South

and S.E. Before the pampero bursts forth the barometer is very low; the mercury begins rising towards the end of the squalls, when the wind shifts to South.

In the River Plata, and also at sea in the same parallel, the winds are very changeable: during the fine season, from September to March, the prevailing wind out at sea is from N.E.; the horizon is charged with vapour, and the sky filled with clouds of indefinite forms. On nearing the river the wind veers to East, sometimes to S.E., very fresh, with rain and cloudy weather.

In the interior of the river, during this season, the wind from S.E. blows regularly in the afternoon; at night it falls and shifts to North: this wind is called a *virazon:* when it falls, and the wind from North to N.W. continues, a storm from S.W. *(pampero)* may be expected, more or less violent according to the *virazon.* We repeat, that great precautions are necessary against these storms; they may prove fatal to those who are not prepared to meet them.

About the times of full and change of the moon, strong breezes are found from S.E., with rain; sometimes also the wind blows from North, not so strong as that from S.E., and the temperature is higher.

The pilots say that the S.E. wind blows when the moon has South declination, and the North wind when she has North declination. In these cases the North wind generally shifts to N.E. in dry weather; if accompanied by rain or heavy dew it veers to N.W. It often becomes violent, blows in squalls from this direction, and ends by shifting to S.W., blowing strongly; with this wind the sea rises suddenly, and subsides as soon as it ceases.

From March to September the general winds at the entrance of the Plata are from West to S.W. Ascending the river they are more frequently northward instead of southward of West.

The winter season is preferable to that of summer in the roadstead of Buenos Ayres; for the wind being generally from S.W. to N.W. the sea is smooth and communication facilitated.

In the months of July, August, and September there is frequently a thick fog from the mouth of the river to the shore of Ortiz; further up this is not so frequent.

The inhabitants of La Plata attribute these prejudicial influences to the North wind: it is in fact very hot, and while it blows the air is charged with electricity: thus the wind from this quarter almost always terminates in a storm, during which it shifts to S.W. and restores the equilibrium.

Sometimes the pamperos extend out to sea, and pass the latitude of the island of St. Catherine. When it is clear they last longer than when the clouds are charged. What has been said respecting the winds of La Plata at sea, at its mouth, and in the interior, is what takes place in a general way; but the contrary must not occasion surprise, for the wind is so variable that neither its duration nor direction can be depended on with certainty: frequently during consecutive years at the same seasons the winds are widely different.

East Coast of Patagonia.—Ships leaving the Atlantic and bound for any port in the Pacific, will derive advantage from keeping at least a hundred miles from the East coast of Patagonia, as much to avoid the heavy sea caused by the westerly breezes, which predominate in the East, and are stronger according to the distance from shore, as to profit by the variableness of the westerly wind. Near the coast, from April to September, when the sun has North declination, the wind is more from W.N.W. and N.N.W., than from any other quarter. Easterly winds are very rare, but when they do take place, as they come obliquely to the coast, there is no danger in keeping in shore.

During the opposite season, when the sun has South declination, the wind is principally from the southward of West, and sometimes very strong; but as the coast is to windward the sea falls with the wind. Although during this season the wind may not be fair, yet as it is rarely steady, and often varies six or eight points backwards or forwards in a few hours, advantage may be taken of this circumstance by keeping near the shore.

Terra del Fuego.—Fogs are very rare on the coast of Terra del Fuego; but dark and rainy weather, accompanied by violent winds, is generally experienced there. The sun appears but seldom, and even in fine weather the sky is dull and cloudy, and the atmosphere very seldom clear.

Different winds succeed each other at short intervals and last several days; sometimes the weather is fine for the space of a fortnight; but this happens very rarely.

The equinoctial months are the worst of the whole year about Terra del Fuego and Cape Horn. The winds are then strong, but may not always be expected on the exact day of the equinox.

The months of August, September, October, and November, are also generally worse than the others. During these months westerly winds prevail, as well as snow, rain, and intense cold. December, January, and February, are the hottest months. The days are then long, and the weather sometimes fine. But in these months westerly winds prevail, sometimes very strong, and accompanied by much rain; for even the summer, in these latitudes, only possesses the advantage of longer days and a less rigorous temperature.

March is subject to storms, and is perhaps the worst month in the year, on account of the sudden squalls which then take place. It is, however, not so rainy as the summer months.

In April, May, and June, fine weather is general, and although the days shorten at this time, the weather resembles that of summer more than at any other period of the year. Bad weather is nevertheless found during these months; but the easterly winds, which are frequent, bring with them some fine days.

Passage from the Atlantic to the Pacific.—June and July are much alike, only in July the easterly winds are more frequent. The shortness of the days and the extreme cold render these months very unpleasant, although they are perhaps the most favourable for sailing from the Atlantic to the Pacific Ocean, because the wind so frequently blows from the East.

On the contrary, the summer months, namely, December and January, are the best for passing from the

Pacific to the Atlantic Ocean, though this passage is so short and easy that it may be attempted at all times.

In these regions thunder and lightning are little known. Violent storms are announced by heavy clouds coming from South and S.W.; they are accompanied sometimes by snow and hail of large dimensions, which render them still more formidable.

Westerly winds prevail during a great part of the year in these parts, and those about Cape Horn. The easterly winds only blow during the winter months now and then: they are however violent during this season, and are very rare during summer.

The easterly winds are always moderate, and accompanied by fine weather; when they begin blowing they gradually increase. The weather then changes, and the breeze perhaps becomes somewhat fresher: they often attain such a force as to require three reefs in the topsails; then they gradually fall or shift to another quarter.

The North winds commence by blowing moderately, only the weather is more gloomy and cloudy than with an easterly wind; and a little rain generally falls. In proportion as the wind freshens, it veers westward, and increases in force, blowing between North and N.W. The sky then is very cloudy and dark, and rain falls abundantly. From N.W. it blows hard, and when its force is expended, (which it is in about twelve or fifteen hours,) or even while it blows hard from this quarter, the breeze suddenly shifts to S.W., and blows more violently than before. The wind disperses the clouds, and in a few hours the weather is perfectly clear; nevertheless, at times the squalls are very severe. The wind continues several days in the S.W. quarter, generally blowing very fresh; then it moderates a little, and, after two or three days, the weather becomes fine.

The North winds generally blow during summer, and it is a fact established from observation that the shifting of the wind from North to South takes place by the West during this season,—one which would little deserve its name were not the days longer and the atmosphere warmer. The winds and rain are much more violent during the long than during the short days.

It should not be forgotten that bad weather never comes suddenly from East, and that a S.W. or South wind never shifts quickly to North. On the contrary, winds from South and S.W. come on suddenly and with violence.

South winds and storms from S.W. are preceded and announced by thick masses of large white clouds rising in these parts, the borders of which are clearly defined, and which appear round and solid.

The North winds are preceded and accompanied by very low clouds, the sky is overcast, and some clouds appear above the rest. The sun can scarcely penetrate them and it assumes a reddish aspect.

Some hours and even a whole day before a North or N.W. wind the altitude of the sun cannot be taken, although it is visible, because the mist of the atmosphere prevents its disc from being distinctly seen.

Sometimes, but rarely, with a slight breeze, varying from N.N.W. to N.N.E. there are a few days of fine weather; South breezes and rain most generally succeed. The most common weather in the regions mentioned above, is a cloudy sky with a fresh breeze, varying from N.W. to S.W.

Falkland Isles.—It would be difficult to find a region more exposed to storms in summer and winter than the Falkland Islands.

The winds there are very variable, rarely falling while the sun is above the horizon, and sometimes very violent, even in summer. A day of calm is an extraordinary fact at the Falkland Isles. Generally, it blows less during the night than during the day; but, both by day and night, at all times of the year, they are exposed to sudden and violent storms and squalls of wind, though they do not, usually, last more than a few hours.

The prevalent wind is westerly. It generally begins from N.W., shifting to S.W. by West; and when the N.W. wind is attended by rain it quickly passes to S.W. and blows with strength from that quarter.

The North winds produce cloudy weather, and when they are light are accompanied by thick fog. It is also observed that they usually blow about the times of the moon's quartering. The winds from N.E. and North produce very gloomy weather, with much rain. They are sometimes strong and veer to N.N.W., but most frequently to West. S.E. winds also bring rain. They are rather frequent and blow strong, and in proportion as they strengthen they veer to the southward.

During winter the principal winds are from N.W.; and in summer from S.W. Although sometimes fogs attend the wind from East and North, they do not often last for a day.

The squalls of wind from the South, from S.W., and S.E. are more violent and sudden than those from any other direction. East winds are seldom strong, and last only a short time. They generally produce fair weather and may be expected more during April, May, June, and July than at any other time of the year. Intervals of fine weather are very scarce in the course of the year when the wind is varying between E.S.E. and E.N.E.

Thunder and lightning are very rare, and with the latter easterly winds may be expected. If lightning appears in the S.E., and the barometer is low, a heavy breeze of wind from that quarter will most likely follow.

These breezes from S.E. and South last longer than those from West, at least generally, and they cause a heavy surf on the South coasts of these islands. In winter the wind is generally not so strong as in summer, and during the former season the weather, though colder, is drier and better established.

All important changes are foretold by the barometer, provided the changes of the mercury are understood by those who consult it, and frequent observation is made.

Such are the general observations on the winds in the different parts of the basin of the Atlantic. We shall next proceed to consider the currents of this sea.

GENERAL VIEW OF THE CURRENTS OF THE ATLANTIC OCEAN.

The currents of the Atlantic are of two kinds: one occasioned by tides, and observed only at short distances from the coast; the other by causes differently explained; but almost constant in their direction, and deviating only near those coasts which impede their progress.

These last, called general currents, are divided into cold and warm currents, according to the waters of which they are composed. The result of all observations on currents may be thus expressed:—cold currents flow from the poles towards the equator on the western coasts of continents. Currents setting from East to West flow along the equator; warm currents flowing from the equator towards the poles pass along the eastern shores of great continents. Thus we find a cold current setting from North to South on the western coasts of Europe and the N.W. coast of Africa; a cold current from South to North on the S.W. coast of the same continent. But, on the contrary, on the coast of Brazil a warm current is found flowing from North to South; then a warm current, which after having circulated through the gulf of Mexico leaves it by the Bahamas, and flows along the coast of the United States, being known by the name of the Gulf Stream.

Such are the general currents of that vast basin, called the Atlantic Ocean, formed by an immense longitudinal valley, separating the European and African continents from that of America. At present we will confine ourselves to the notice of these currents.

Philosophers differ as to the originating causes of these general currents. Some attribute them to the action of the trade winds; while the greater number admit that, like the winds, they are produced by the sun's heat, and by the rotatory motion of the earth. Thus, they say, in consequence of this movement, and the passage of the polar waters towards the equator, a current, apparently directed from East to West, must be formed at the equator, according to the same that takes place in the atmosphere and produced by the same cause.

Now, a constant current must necessarily produce a drain of the adjacent waters towards one of its extremes, say the eastern, and on the contrary a lateral displacement at the opposite one, the western; or, in other words, admit a flow of the polar waters towards the equator on the western coasts of the great equatorial continents, then on the contrary a flow towards the poles on the eastern sides of the continents must ensue. We may further observe that the equatorial waters as they flow onwards for the poles, on account of the greater velocity of rotation at the equator, as well as on account of the flow of the polar waters towards the equator, should follow a certain direction easterly from the West, like the Gulf Stream in the North Atlantic and the current of the South Atlantic flowing from the coast of Brazil towards that of Africa.

Equatorial, Polar, and Tropical Current.—We call

the equatorial current that which flows from East to West at the equator. We shall distinguish by the name of *polar currents* those flowing from the poles towards the equator on the western coasts of the continents; and by the name of *tropical currents* those flowing from the equator towards the poles on the eastern coasts of the continents. The velocity of these currents varies in different parts of their courses; the greatest that has been observed is from 60 to 120 miles in twenty-four hours. Their general temperature is higher or lower than that of the sea through which they pass, according to that of the climate where they originate. We shall now allude to the direction and limits of the Atlantic currents, showing their mean velocities and temperatures; and first those of the equatorial current.

Equatorial Current.—The equatorial current commences on the West coast of Africa in about 5° 30′ E. of Greenwich. It passes by the isle of Anno Bom and continues westward parallel to the equator between 1° and 1° 30′ N. lat. and 2° or 3° S. It soon extends itself northward and southward, and although it hardly passes North of the parallel of latitude abovementioned, it ranges abreast of Cape Palmas between 3° 30′ and 5° S. lat. For the space of a thousand miles, nearly as far as 12° or 14° W., along the equator in the gulf of Guinea it runs by the side of another current taking the opposite direction from East to West, called the Guinea Current. This part of the sea then presents the remarkable phenomenon of two currents adjacent to each other running with great velocity in opposite directions, and having a difference of temperature of about seven degrees; so that imagining a vessel to be in either of these currents sailing eastward in the gulf of Guinea, her progress would be accelerated or retarded forty or fifty miles a day, that being the rapidity of the two currents in this part. We shall hereafter return to this important fact.

Advancing westward on each side of the equator, the equatorial current in 20° or 21° W. throws itself into the northern hemisphere under the name of the N.W. branch of the equatorial current, reaching to 20° N. lat., extending itself gradually, and sometimes is even felt as far as 30° N. Then again it also extends more southerly, and thus covers a space of three hundred miles, nearly reaching to Cape St. Roque, where it becomes divided into two distinct branches. The northern branch, which is the largest, forms the Guiana Current, and reaches to the Antilles; the southern branch, running parallel to the coast of South America, and extending far out to sea, forms the Brazil Current.

The Equatorial Current—Extent.—The length of the equatorial current from the coast of Africa to Cape St. Roque is 2,500 miles, and to the Antilles 4,000 miles. Its breadth near its origin is 160 miles; abreast of Cape Palmas 360 miles; and it runs for 450 miles before it divides.

Velocity.—This current has most velocity in summer and least in winter. Between the meridians of 5° E. and 8° W. long. its mean rate is from 25 to 30 miles a day. Between 8° and 14° W. long., towards the end of June and the beginning of July, it varies from 44 to 75 miles. Between 14° and 21° W. long., from 45 to 60;

its mean velocity may then be estimated at 46 miles in twenty-four hours.

Temperature.—The mean temperature of its waters is 73°, or about 4° or 5° above that of the ocean in different seasons.

The N.W. branch of the equatorial current, flowing first to N.W., assumes afterwards a more northerly direction. This current is always felt in 18° N. lat., and sometimes even as far as 30°. It is lost in the current produced by the N.E. trade, to which it appears to give a N.W. direction, besides increasing its rapidity. The width of this current at the point of separation is nearly 200 miles, and further North 300 miles. Its velocity up to 10° N. lat. is 20 or 24 miles a day; from that it gradually decreases. This current is important to ships passing to the northward from the southern hemisphere.

The general set of the waters towards the West in the zone adjacent to the equator, here alluded to, cannot be doubted. Nevertheless, several navigators have asserted that in the same zone, and for several days, they have experienced currents flowing to East. Some commanders have crossed the easterly current setting at a rate varying from 0·3′ to 1·0′ per hour.

These cases, we conceive, may be regarded as exceptions, and the currents as eddies or counter-currents on the limits of the general current. Some authors, however, mention the existence of a current running from West to East between 8° and 10° N. lat. It is felt first about 53° W. long., and will reach as far as 26° W. On the first meridian its general direction is between North and N.N.E., and in proportion as it advances towards East it increases in breadth. It also increases in velocity till it reaches to about 38° W. long., and its direction becomes more easterly. Beyond the meridian of 38° its velocity decreases, while it becomes more extended, and it can hardly reach the longitude of 26° W. We recommend this current to the examination of seamen.

We have hitherto alluded to secondary currents, which philosophers attribute to winds blowing in the same direction for a longer or shorter period. The secondary currents, which owe their origin to winds as constant as the trade, are constant themselves, and flow in the same direction with nearly uniform strength. They are found in the Atlantic between the tropics, but are only considered regular between 23° N. lat. and 9° S. lat., the space in which the trade winds blow regularly, and they attain a mean velocity of nine or ten miles per day. The currents caused by the prevailing winds are neither constant in their direction nor velocity; they are found both North and South of the 32nd degree of latitude.

The Guiana Current.—The Guiana Current, which is a continuation of the equatorial current, runs along the low coast of Guiana towards the isle of Trinidad. About the equator it is crossed by the waters of the Amazon, a river which, receiving an immense volume of water from tributaries, forms a counter-current to it, producing considerable overfalls. This, however, owing to the impetuosity of the fresh water, does not influence its direction. The river waters and those of the Guiana current do not intermingle with each other; for after crossing that current the river water is recognised at 300 miles from its mouth.

A little South of Trinidad the River Orinoco discharges a considerable quantity of water into the equatorial current. From the nearly similar directions in which they run the waters easily combine with each other, and the rapidity of the current is thus considerably augmented. It then enters the sea of the Antilles by the strait formed on one side by the isle of Trinidad and on the other by that of Martinique; in which space are the islands of St. Vincent, St. Lucia, Grenada, Barbados, and Tobago.

Current of the Carribbean Sea and Gulf of Mexico.—In the Carribbean Sea no constant currents have been observed; and although in the midst of this sea and about the islands which bound it on the East and North variable currents are found, but generally setting westward, yet along the coast the general current still prevails, following the direction of the coast at a variable distance. Thus, it flows from East to West between the isle of Trinidad and Cape Agulhas; thence it proceeds W.N.W. and N.W. as far as Cape Catoche, crossing the gulfs of Darien and Nicaragua and the bay of Honduras; then it takes a complete circuit of the gulf of Mexico. Thus, after reaching Cape Catoche it turns westward towards the shores of Campeche, along the coast of Yucatan; it thence continues towards Vera Cruz, changes its direction and flows northwards as far as the Rio del Norte and even beyond that river; it flows afterwards N.E. till it meets the waters of the Missisipi; then takes a S.E. direction towards the Tortugas. At this point its direction becomes East, then N.E., and, lastly, North, following the Florida Channel, and discharging a second branch across the Bahama Islands that loses itself in the Atlantic Ocean.

In the middle of the gulf of Mexico the waters do not appear to follow any particular course, as is the case in the Carribbean Sea, and they most frequently depend, as to strength and duration, on the prevailing winds.

The temperature of the sea of this archipelago has been generally found higher than that of the ocean of the same latitude; but, notwithstanding the increase of caloric carried to its water by the surrounding continents, it is considersd that the high temperature of the Carribbean Sea arises in a great measure from the currents of heated water that penetrate it from the Torrid zone of the North Atlantic Ocean.

Velocity of the Guiana Current.—The Guiana current varies in strength during its course (which is about 590 miles) from 10 to 21 or 36 miles a day. It has been sometimes found to be four miles an hour, while near the coast it gradually diminishes to less than half a mile an hour.

Temperature.—The temperature of the waters of this current has been estimated at 80°; that of the waters of the River Amazon very near the line of demarcation is also 80°. The line of separation between the waters of the Amazon and those of this current is N.W.½ N., and the two waters are as distinct from each other as two separate fluids.

Gulf Stream or Florida Current.—The Gulf Stream has its origin in the gulf of Mexico, and the waters having been heated there flow across to the Bahama Channel. Issuing from this strait they flow along the coast of Florida, over 31° N. lat., in a more N.E. direction as far as Cape Hatteras. There, from the indentation of the coast, the West limit of the current takes a more northern direction, while its principal bed is still directed N.E. until it reaches the shoals of St. George and Nantucket, where its direction becomes more easterly. Soon after its direction is E.b.N., passing the southern extremity of the great bank of Newfoundland; and it preserves this direction, between 35° and 43° N. lat., till it reaches the meridian of 36° W. There it turns S.E. and South; afterwards, passing the archipelago of the Azores on the West, it loses itself in the ocean. Its warm waters have, however, been found sometimes on the West coasts of Europe; they have been recognised between the parallels of 44° 20′ and 39°, once also, among others, by Franklin.

Extent and Velocity of the Gulf Stream.—The course of the Gulf Stream is about 3,000 miles from its source to its termination West of the Azores. It traverses in this course nearly twenty degrees of latitude, —from the parallel of 23° to that of 43°.

According to observations made regarding this current, its mean velocity from the entrance of the Florida Channel, at the island of Bemini, to 31° N. lat., is about 70 miles in twenty-four hours. A mean speed of 80 miles a day has been found between the parallels of 26° and 27° N. lat., although a strong North wind blows against it. At the outlet of the Gulf Stream, off Cape Carnaveral, it is like a torrent, and sometimes attains 120 miles in twenty-four hours. It gradually decreases in strength in its progress to the East. Between the meridians of 65° and 66° W. long. it is 55 miles a day; and on that of 42° 30′ it is only 30 or 35 miles. The rapidity of the Gulf Stream afterwards diminishes more rapidly when it curves to the South or to the West of the Azores; near these isles it does not run more than 10 miles a day.

Temperature.—The mean temperature observed in the waters of the Gulf Stream is 86°, which makes it 9° above that of the ocean under the same parallel; ten degrees further North it is found to be 84°, having in this space diminished about 2°; in 61° W. long. it is found to be 81° in summer, and 76° in winter; in 43° W. long. 75°; and in 38° W. long. 73°. Thus the temperature appears to decrease with the rapidity, but not so quickly, as the waters advance eastward; but they still have a very high temperature when they turn towards the South. On coming out of the Bahama Channel, the waters of the Gulf Stream have a blue tinge, and the line of their separation from the waters of the Atlantic is perfectly evident for the space of a hundred miles.

N.E. Branch of the Gulf Stream.—At the place where the Gulf Stream curves towards the S.E., to the northward of the Azores, in 36° W. long., a portion of it continues in a considerable branch towards the N.E., about 600 miles in length, which continues towards the pole, passing between Iceland and the coast of Norway and surrounding the Ferroe Isles. The water of this current is warm, and its temperature in summer has been estimated at 54°; in winter at 51°. Its direction is towards N.E.; but there are very few data as to its rapidity. This current is important to ships bound from the West Indies to Norway, Denmark, or to places situated on the northern coast of the North Sea.

Arctic Current.—It is considered that the arctic current takes its rise in the frozen regions surrounding the North Pole; from whence it descends along the East coast of Greenland towards Cape Farewell. It passes round the cape, a large portion proceeding along the West coast of Greenland till it reaches the latitude of 66° N.; it then turns and again flows towards the South, along the coast of Labrador, forming the current known by the name of the Hudson Bay Current.

In arriving at the North extremity of Newfoundland it sends a branch across the strait of Belle Isle, which mingles with the waters of the St. Lawrence and continues along the South coast of Newfoundland, while the principal current continues down between the Great and Little Banks of Newfoundland and meets the Gulf Stream between 41° and 45° W. long. It there divides into two branches: the one, flowing South towards the sea of Antilles, reaches that part by an under current; an hypothesis which serves to explain the presence of fields of ice met with in crossing the Gulf Stream. The other branch of the arctic current, flowing towards S.W., past the island of Nantucket, forms the counter-current of the United States, occupying the space comprised between the Gulf Stream and the coast extending from Cape Hatteras to that of Florida. The arctic current thus replaces the hot waters of the gulf of Mexico caused by the Gulf Stream.

This current facilitates very much the navigation of the coast of the United States from the northward. It is a cold current, as we have said, and consequently it will be easy to keep in it by means of thermometrical observations, and also to avoid entering the Gulf Stream.

Although to the West of the Azores the current of the Gulf Stream turns partly towards the South, yet between the archipelago and the coast of Europe a general movement of the waters from West to East ensues. This current is known to mariners by the name of the "Bay Current." Can this current be produced by the warm waters of the Gulf Stream, or is it occasioned by the cold waters carried from the pole towards the equator, is a question which has not yet been solved.

The rapidity of the bay current is very variable; it is sometimes 10 or 14 miles a day, and sometimes 24. In the latitude of Cape Finisterre the direction of it is from E.S.E. to S.E., and it divides into two branches; one forming the Rennel Current, the other that of the coast of Portugal.

The Rennel Current.—The Rennel Current, which bears the name of the learned Major who first discovered its course, has an easterly direction near Cape Finisterre. It flows along the North coast of Spain, then proceeds North along the West coast of France, where it is felt at thirty or forty miles off shore, and is fifteen or twenty miles across. It has been found to run from one half to two thirds of a mile per hour. It is very variable, according to the strength of the wind and its direction. It is sometimes found to flow at the rate of a mile an hour, and to this current is attributed the loss of many vessels in the English Channel. It becomes wider as it proceeds northward; and in the latitude of Brest it is eighty miles across, and its direction nearly N.W. It issues from the bay of Biscay, passes West of Ushant at fifteen or twenty miles from that island, crosses the entrance of the Channel, and takes a westerly direction from the Scilly Isles. At the entrance of the Irish Sea it discharges a second branch into that sea, the principal branch flowing W.N.W. and West, towards Cape Clear, and losing itself, near the meridian of 18° W. long. in the polar current towards Northern Africa.

The Portugal Current.—The second branch of the bay current, called the Portugal Current, flows from Cape Finisterre towards the S.S.E. and S.E. along the coast, on which, however, it partly approaches. Off Cape St. Vincent its direction becomes S.E., and proceeding South, it becomes more and more easterly towards the strait of Gibraltar; towards which, under the meridian of 18° W. long., all the waters comprised between Cape St. Vincent on the North, and Cape Cantin on the South, are directed, forming the Strait Current, which carries the waters of the ocean into the Mediranean.

The velocity of the Portugal Current has been found to be from 12 to 24 miles a day. It is very variable, according to the prevailing winds, their strength and duration. This current, then, on the coast must never be trusted, especially in the winter or with strong N.W. winds, when it is necessary to keep well off the coast. The same attention also must be devoted to the current generally of the bay of Biscay, known as the Bay Current, and that of Rennel. These two currents are strong when the West winds, changing from N.W. to S.W., have continued long and with force. In this case it will be prudent to look out for the approaches of the English Channel, and on leaving the bay of Biscay to double Cape Finisterre well out to seaward.

This gulf, then, presents us with this important fact that while out at sea the waters are setting towards East, E.S.E., and S.E., as proved by a number of bottles found near Bayonne and the basin of Arcachon, the waters of the interior and near the coast of France make their escape towards the North and N.W.

Polar Current of Africa and Current of North Guinea.—The polar current of Africa takes its rise in that part of the Atlantic situated abreast of the coast of France. Between Cape Clear, of Ireland, and Cape Finisterre, of Spain, it flows South towards Cape St. Vincent. Between this cape and Cape Cantin of Morocco, the entire mass of water, as far as 18° W. long. flows to the S.E. and N.E. towards the strait of Gibraltar; through which the waters rush like the conduit pipe of a funnel. From Cape Cantin to Cape Bojador, between Madeira and the archipelago of the Canaries, it is directed more easterly and S.E.; but it does not extend in this part and in this direction further out than 150 or 180 miles from the coast. Further out at sea its direction is South and S.S.W. From the archipelago of the Canaries to Cape Verd its direction is generally from South to S.S.W. Afterwards, at Cape Verd, it flows towards South, a little easterly, following the coast of Africa, and takes the name of the North Guinea Current off Cape Palmas.

The western limit of the polar current, near Cape

Verd, is between the island of Sal and that of San Nicolas; afterwards between the island of Mayo and that of Santiago, in the archipelago of the Cape Verd Islands. Its direction is from South to S.W. nearly all throughout its course from its rising till it reaches this part. Off Cape Mesurado its direction becomes E.S.E., and even East out at sea; while at a little distance from the coast it is S.E. as far as Cape Palmas. Off this cape its direction out at sea is easterly; then E.N.E. as far as the gulfs of Benin and Biafra. It then meets with the equatorial current, and after having reached Princes Island the waters probably mingle with those of the equatorial current.

Extent of this Current—The width of the North Guinea Current varies according to the seasons. In the latitude of Cape Palmas it extends nearly 180 miles to seaward, that is, as far as 12° W. long.; and it occupies the space comprised between the parallel of 2° 30′ N. lat. and the North coast of Guinea. On the meridian of Cape Palmas it is nearly 150 miles across; but to the East, in the gulf of Benin, it attains to a considerable breadth, nearly 300 miles from North to South. It is not felt in the vicinity of Isle St. Thomas, neither is the equatorial current, which is only first found a little to the West of this island, in about 6° E. long.

Velocity of the Polar Current of North Africa.—The velocity of the polar current of North Africa near its origin on the coast of Portugal, is about 12 miles a day. On the coast of Africa it varies from 16 to 10 miles till it reaches Cape Palmas.

Velocity of the North Guinea Current.—The current of North Guinea flows with the greatest rapidity from June to September. When East of Cape Palmas this is found to be 40 or 50 miles a day. Off Cape Three Points it is nearly 34 miles per day. It then decreases, and in the gulf of Benin its direction is from East towards the South.

Temperature of the Current.—Near the Cape Verd Islands, the temperature of the waters of this current is 8° or 10° below that of the waters of the ocean; it then increases rapidly in proportion as it proceeds South. In the gulf of Guinea the temperature of the water has been observed to be 84° in the middle of the current; 83° and 81° at its southern limit; in contact with the colder waters of the equatorial current it is 79° or 81° in the North part, adjoining the coast. This current is of the utmost importance in navigating the western coast of Africa.

Such are the general currents of the North Atlantic Ocean, the other portions of which are occupied by currents flowing from these; the principal of which, as above stated, is that flowing towards the West and S.W., caused by the constant trade winds from N.E.

North Atlantic Ocean—Comparative Table of the mean velocity of the Currents for twenty-four hours.

Equatorial Current	46 miles.
Guiana Current	30 ,,
Gulf Stream	33 ,,
Current caused by N.E. trade winds	10 ,,
Rennel Current	18 ,,
African Current and North Guinea	20 ,,

Brazil Current.—We have already spoken of the

current of Brazil, a southern branch of the equatorial current, dividing at Cape St. Roque. It extends 250 or 300 miles along the coast of South America, and commences from 6° or 7° S.

The space between the coast and this current is occupied by other currents which follow the direction assigned to them by the alternate S.E. and N.E. winds of the coast of Brazil. The current of Brazil is crossed by the waters of the River Plata, which may be recognized more than 200 miles from the mouth of that river. These waters do not, however, appear to produce much effect on the Brazil Current, which in these latitudes seems to divide into two branches. The most considerable, taking an easterly direction, forms the counter-current of the South Atlantic Ocean. The other branch, flowing southward, forms a current which, though very feeble, is sometimes felt as far as the entrance of the strait of Magellan. The mean velocity of this current in the part nearest the equator, is about 20 miles a day.

Alternate Currents of the Coast of Brazil.—We have said that between the coast of Brazil and the current of which we have spoken alternate currents are met with, occasioned by the periodical winds which blow on this coast. The force of these currents depends on the strength of the wind, and consequently is very variable. From March to September, when winds from S.E. to E.S.E. prevail, the current sets northward, and from September to March, with N.E. winds, veering to E.N.E., it sets southward; but these directions are much varied by the form of the coasts. This current is felt only about 50 or 60 leagues from the coast of Brazil, and is of the utmost importance to navigation.

Current of Cape Horn.—The current off Cape Horn sets constantly from the Antarctic Sea and round the cape from the Pacific Ocean into the Atlantic, and is generally accompanied by strong westerly gales. Its general direction is E.N.E. and N.E. However constant may be the prevailing winds on the East coast of America, it flows to N.E., passing the Falkland Isles. In some seasons it preserves its N.E. direction as far as the parallel of 49° or 48° S. lat., and it is most probable that it joins the counter-current of the South Atlantic Ocean, of which we shall speak presently.

On the coasts of Terra del Fuego the rate of this current has been found to be 12 and 15 miles in the course of the day. In 57° S. lat. and 72° W. long. it is 35 miles. Near the coast its mean rate is about 24 miles. While between Cape Horn and Staten Land, in 55° S. lat., its direction is N. 51° E., and its rate 56 miles per day.

The waters of this current, flowing northward and partly coming from the Antartic Polar Sea, have a lower temperature than those of the adjacent ocean.

That part of the South Atlantic occupied by the counter-current which flows from the coast of Brazil towards the Cape of Good Hope, is only partially known. It is considered in a great measure to be formed by the tropical current of the coast of Brazil; but this is not established as certain. It flows rapidly to the eastward, passing 150 or 180 miles South of the Cape of Good Hope. It then penetrates the Indian Ocean, and traces of it are found more than 2,000 miles beyond the Cape, where it unites with the polar current of Australia. This current is very favourable to ships rounding the Cape to the eastward.

Current of the Cape of Good Hope.—The current off the Cape of Good Hope is formed of two others from the Indian Ocean: the principal of which flows southward from the Mozambique Channel along the African coast; the other coming from that part of the ocean southward of Madagascar, is the S.W. branch of the equatorial current of the Indian Ocean. These two Indian currents unite a little South of Port Natal, where they take a more southerly direction over the bank of Agulhas. Instead of then flowing entirely, as one might imagine, into the Atlantic Ocean, the greater portion of this returns by a counter-current into the Indian Ocean, mixing with the counter-current of the Atlantic, abovementioned, after the Agulhas Bank has modified the direction of the current off the Cape of Good Hope to the West. This current is known as the counter-current of the Cape of Good Hope.

South Atlantic Polar Current.—The other portion of the current off the Cape of Good Hope flows into the South Atlantic Ocean, following nearly all the western coast of Africa. This branch is called the polar current of the South Atlantic, or the South Atlantic Current.

At the point of junction of these two currents, near Cape Recif, the current of the Cape of Good Hope is 90 or 100 miles wide. In some parts its velocity is from 60 to 100 miles in twenty-four hours.

Temperature.—Beyond the bank off the Cape its temperature has been marked at 70°, that is, 9° above that of the ocean. Near the edge of the bank it is found to be 68°, namely 7° above the temperature of the ocean. On the bank itself the temperature of the water has been found 5° below that of the ocean.

The counter-current of the Cape of Good Hope, above-mentioned, is sometimes 210 and even 240 miles wide.

South Atlantic Polar Current.—The polar South Atlantic current, which, as we have seen above, is the only branch of the current off the Cape of Good Hope which penetrates the Atlantic Ocean, flows round the Cape of Good Hope, extending 80 or 100 miles out to sea, passing to the northward of the counter-current of the Atlantic Ocean. It increases off the Cape even to about 150 or 160 miles in extent. It then flows northward, enlarging more and more, when its principal direction becomes N.W. Near the Cape of Good Hope, and nearly in the same latitude, it receives a branch which appears to proceed from the counter-current of the Atlantic Ocean.

Velocity.—The rate of this current has been found generally 16 miles in twenty hours, in a N.W. direction generally. In its course it meets with the waters of the River Congo, flowing with great rapidity; but they do not appear to attain any decided influence over it, the direction of it and those of the current forming no considerable angle. The waters of the Amazon do not mingle with the ocean for a considerable distance, and at 200 miles from the river's mouth the more highly coloured water of the Congo may be perceived.

Temperature.—Near the Cape of Good Hope the temperature of this current is 67°, that is, 3° above that of the ocean. On the parallel of 30° S. lat. it is not more than 64°.

Current of South Guinea.—From St. Paul de Loando the current continues along the coast of Africa, flowing nearly N.W. Off Cape Lopez a portion of the principal current takes a more northerly direction, following the coast of Gaboon as far as the gulf of Biafra. It is there lost in the equatorial current and may be called the current of South Guinea. The principal direction it takes is N.N.E. and N.E. near the coast, and N.W. further at sea and near the isles of the gulf of Biafra. At sea the limit of this current appears to be to the eastward of Princes Island.

Velocity.—The velocity of it is often 24 miles per day, but is generally about 10; it is, however, very variable, sometimes ceasing altogether. Still, southerly currents are found in this part of the gulf of Guinea, but the circumstance is very rare.

Temperature.—The temperature of the current of South Guinea at its limit near the coast of Gaboon, has been found to be 77°, and in the middle of the bed of the current 77° and 79°. Thus it is 6° or 8° higher than that of the waters of the equatorial current. The knowledge of this current is very useful to vessels sailing towards Gaboon.

The vast portion of sea forming the centre of the South Atlantic, is occupied by currents produced by the S.E. trade wind: their general direction varies from West to S.W. and S.S.W. in proportion as the waters approach the exterior limit of the current of Brazil, with which they mingle in order to return eastward by the cross current of the Atlantic.

South Atlantic Ocean—Comparative Table of the mean velocity of the Currents for twenty-four hours.

South Atlantic Current	15 miles.
Brazil Current	20 ,,
Counter-current of the Atlantic	15 ,,
Current of Cape Agulhas	80 ,,
Counter-current of the Cape	30 ,,
Current caused by S.E. trade	10 ,,
Current of South Guinea	10 ,,

GENERAL REMARKS ON THE NAVIGATION OF THE ATLANTIC.

It is a general rule in the navigation of the ocean when going from East to West to attain, if convenient, the zone of the trade winds; and to avoid it when going in the opposite direction. In the first case, then, it becomes desirable to reach it, and in the other to leave it, as soon as possible.

Having treated on the prevailing winds and currents of the Atlantic Ocean, we shall now allude to the routes which should be taken for crossing it.

Routes from Europe to North America.—In these routes from Europe to North America, it is generally acknowledged that the further North the port of departure is the greater are the chances of a speedy passage.

In the beginning of the year it is advisable to keep North of 46° or 47° N. lat. as far as the meridian of about 32° W., and then to haul South to the parallel of 43° N., and keep in or near this parallel without making northing, especially in approaching the coast of North America, in order to pass well clear of Sable Island, this being so dangerous that it cannot be avoided too carefully. By following this route the northern limit of the Gulf Stream will be avoided, and after leaving Newfoundland the arctic current will assist in the track to the S.W. for the ports of Nova Scotia and New Brunswick, or those of the North United States.

Towards the end of the year it may be better to adopt a course to the northward of that. Thus, leaving Europe, proceed to the N.W. as far as 55° latitude and 30° W. longitude. From thence cross the banks of Newfoundland on a S.W. course in 46° latitude; then pass about sixty miles South of Sable Island, and from thence make for the desired port.

In these passages it is recommended never to pass northward of Sable Island, on account of the frequent fogs met with in those regions and strong S.W. currents that are found near it, the effects of which cannot be foreseen.

Routes from Europe to the Ports of the United States.—Passages from Europe to the United States are much retarded by the Gulf Stream, which should be avoided, for in case of contrary winds or calms an easterly set would be inevitable. In order to reach these ports, then, the routes previously indicated should be followed, passing southward of Sable Island, and from thence following in the southerly current which flows along the coast of the United States, in order to avoid that of the Gulf Stream. In all cases if this current is to be crossed to the westward it should be done as quickly as possible.

There is another route which, although longer as to distance, appears preferable; for if the time occupied in the passage might appear greater in consequence of the distance it is really less as to the speed with which the vessel would sail from port to port. This route is that of the trade winds. On leaving Europe, if the wind be not favourable to a direct route towards the ports of the

United States, it would be better to make good a course South or S.W., as the wind permits, in order to find the trade winds as quickly as possible. The best course to reach their latitude is either between the Azores and Madeira or Madeira and the Canaries. It would be better to avoid passing between these last named islands and the coast of Africa, because the trade wind there loses its force and direction. But a vessel when once in the region of the trade winds may pursue the most convenient course, according to her desired port, only being cautious as to making the land, and in crossing the Gulf Stream, so as to be about ten leagues or so to windward of her port.

There are, however, many circumstances under which this route can be made without the assistance of the trade winds, and they occur principally during the forty or fifty days after the two equinoxes, periods in which N.E. winds are frequently found; so that vessels sailing then may shape their course at once. Besides, if a vessel in the counter current of the Atlantic meet with contrary winds, it is better to make southing, in order to fall in with the trade, than to be striving against these winds. In the spring, summer, and autumn seasons, when the N.E. trade winds extend as far as 28° and 30° N. lat., the passage by the trade winds will be advantageous. Lastly, if the wind admits of it when going from Europe to the United States, West is the course to adopt; if not, and if at the time of the equinoxes, adopt that which is the nearest to it. In any other case we should prefer adopting a southern course, so as to attain the region of the trade winds.

On comparing the passages of ships made during six years between Liverpool and New York, it appears that the passage is made on an average in forty days. As a specimen of quick passages we may mention the following: the *Charlotte*, sailing from Bremen to New York, has made two voyages in thirty-three and twenty-eight days; the *Alexander*, starting from the Weser, has been twenty-seven days in reaching the same port; and the *Clementina*, starting from Bremen, has reached Baltimore in twenty-nine days.

Homeward Course from the United States to Europe.—In the homeward course from the ports of the United States to Europe, those currents which set to the southward should be crossed as quickly as possible, so as to gain the Gulf Stream and attain a northern latitude in order to get clear of this current, because it is frequently subject to bad weather; and in the months of July, August, September, and October severe weather is experienced in it. During the other months, however, probably a good vessel might keep in it, and would thereby much shorten her passage. When on the meridian of 42° W., the course should be directed so as to pass to the northward of the Azores; and from thence, according to the winds, to follow the course most convenient for reaching the port of destination. These passages are greatly assisted by West winds, veering to S.W. and N.W. In fact, the general passage made by sailing packets from New York to Liverpool, deduced from all the voyages made by them during six years, is twenty-three days. The same passages made by steamers present the following results:—

From East to West.	*Longest.*	*Shortest.*
Great Western, Bristol to New York.........	21½ days.	13 days.
Royal William, Liverpool to New York ...	21½ „	18½ „
Liverpool, Liverpool to New York	18½ „	16 „
British Queen, Portsmouth to New York...	20½ „	14 „
From West to East.		
Great Western, New York to Bristol.........	15 „	12 „
Royal William, New York to Liverpool ...	17½ „	14½ „
Liverpool, New York to Liverpool	17½ „	13¾ „
British Queen, New York to Portsmouth...	22½ „	13½ „

In leaving Europe for the Gulf of Mexico or for ports of the Carribbean Sea, as soon as an offing is obtained the course should be S.W., in order to reach the region of the N.E. trade winds as soon as possible. In this part of the route care must be taken not to approach too near the coast of Africa, in consequence of the current and the wind becoming more westerly. If obliged to continue as far South as the Canaries to find the trade they should be left to the eastward.

A vessel once in the region of the trade winds, bound to the Lesser Antilles, may make directly for her port, keeping as long as possible on the parallel of 19° or 20° North latitude, from the month of May to December. From December to June, on the contrary, a more southern track should be followed. But in approaching the Antilles much allowance must be made for the current, as the reckoning will always place the vessel East of her true position. In such cases it will be well to add twelve miles a day to the westward course to allow for this current. If the vessel be destined for the Great Antilles or the ports of the gulf of Mexico she will enter the Carribbean Sea between Guadaloupe and Antigua, or between Isle St. Martin and Culebra. This is invariably the entrance chosen in voyages to St. Thomas, Porto Rico, Kingston, Havana, Tampico, Vera Cruz, and New Orleans. When bound to La Guayra, Porto Bello, Cartagena, or any of the ports of Venezuela, vessels generally pass between St. Lucia and St. Vincent. Vessels bound for Guiana should keep in shore and to the South of their destined port on account of the currents.

Routes from Europe to Guiana.—On leaving Europe for Guiana, the general route, from November to July, will be to cross the parallel of 10° N. lat. in the most direct line between the meridians of 48° and 50° W., in order to cross the zone of calms to the West of the most difficult part. Having reached the parallel of 10° they would keep a point or a point and a half further South to meet the effect of the general current setting N.W., so as to attain, at about fifty leagues from land, the parallel of 3° or 3° 30′ N. latitude. A westerly course, until in about eight or ten fathoms, might be adopted for the coast. From July to November the following course might be better adopted and sometimes with advantage. Passing 150° leagues to the West of the Cape Verd Islands, steer South, so as to cross the zone of the variables, and reach the S.E. trade, which at this season is felt as far as 5° and 6° or even 7° or 8° N. latitude. Having found the S.E. trade, a westerly course between the equator and 3° 30′ N. lat. would make the coast in a depth of six or eight fathoms.

Vessels from the Lesser Antilles bound to Europe generally pass between Guadaloupe and Montserrat.

From thence, with East and N.E. winds, it is best to make northing in order to get clear of the trade winds as soon as possible. When the zone of the variable winds is attained, a ship should proceed as previously directed in the homeward routes from North America to Europe. Vessels from Jamaica generally pass between St. Domingo and Cuba, and thence between Inagua and Crooked Island. If bound to the Lesser Antilles a vessel should steer between the North coast of St. Domingo and the S.E. shore of the Bahamas. From thence, avoiding the wind, she could reach the Lesser Antilles sooner than by plying to windward in the Carribbean Sea. Vessels from La Guayra, Porto Bello, or Cumana for Europe leave the Carribbean Sea by the Mona Passage, formed by the isles of St. Domingo and Porto Rico. From thence they proceed to the N.E., in order to cross the parallel of 40° N. latitude between the meridians of 30° and 35° W. longitude.

Ships leaving Porto Rico proceed directly North, in order to pass the region of the trade winds, following nearly the same route. On leaving Cuba or the ports in the gulf of Mexico, vessels pass up the Bahama Channel and thence steer to N.E. to leave the stream. They then proceed eastward, passing South of the Bermudas, and again cross the Gulf Stream in the neighbourhood of the Azores.

Vessels leaving certain ports of Costa Firma for the Lesser Antilles, will perhaps derive advantage from adopting the Bahama Channel instead of contending against the wind in the Carribbean Sea.

Outward Voyage.

Elbe to Havana	59 days
Hamburg to Guayra	50 „
Channel to St. Domingo	46 „
Channel to Vera Cruz	40 „
Channel to Antigua	27 „

Homeward Voyage.

Havana to Elbe	49 „
Jamaica to Channel	32 „
Havana to Gibraltar	47 „
Vera Cruz to London	42 „
Guadaloupe to Channel	33 „
Port Prince to Channel	30 „
St. Thomas to Hamburg	45 „

Routes from Europe to South America.—Vessels leaving Europe for the ports of South America, such as Rio Janeiro or Buenos Ayres, ought to steer about South-West to fall in with the N.E. trade wind as soon as possible, passing between the Azores and Madeira, or between Madeira and the Canaries, and to the West of this archipelago, unless required to stop here. Thence they would proceed to cross the line, traversing the zone of the variable winds.

Crossing the Line.—It has been for some time the rule to cross the line in 22° or 25° W. longitude. Numerous facts have proved it preferable to cross it between 25° and 30° W. In fact, between these meridians the zone of the variable winds of the equator is less extended than it is towards the coast of Africa, and it is frequently passed without experiencing calms from the N.E. and S.E. trade winds.

Their change is frequently accompanied by stormy weather. As to the fear of being drawn towards the West and towards Cape St. Roque by the equatorial current, it would seem that this has been much exaggerated and also that the trade winds in this part blow more from the East than they were supposed to do, so that Cape St. Roque may be doubled without any difficulty. As a general rule it may be stated that the winds from the sea on the coast of Brazil blow nearly always at right angles to the line of the coast, principally from October to March. During this period, then, the coast may be approached without fear, the winds being generally from N.E. to E.N.E., and the current near the coast setting from North to South thereby, as observed, assisting the passage. From March to October, on the contrary, the winds coming from East to E.S.E. and the current near the coast setting from the southward, it will be preferable to keep forty or fifty leagues from the coast, in the current of the coast of Brazil, and pass westward of Trinidad in order to reach Rio Janeiro and Buenos Ayres.

Vessels bound to the Pacific by Cape Horn, whether sailing from Rio Janeiro or Buenos Ayres, or coming from the northward, should keep at a distance of 100 miles from the coast of Patagonia, in order to avoid a high sea, caused by the West winds which prevail there, and to profit by the changes of the wind on the coast. They will then pass between the Falkland Islands and Terra del Fuego—that is if they are bound for the Pacific—and will generally pass East of Staten Land, the strait of Lemaire being often difficult to adopt. The courses to be taken in leaving the ports of South America differ according to the latitude of these ports.

Homeward Voyage from South America to Europe.—Vessels from the ports of Brazil to the northward of the point of Olinda may generally stand along the coast on the starboard tack and direct to the northward. Those leaving any port of Brazil to the southward of that point are generally obliged to get on the port tack, to avoid the coast and make a board to the southward. Sometimes the N.E. winds oblige them to continue on this tack for twelve or fourteen days, and standing to the S.E. and S.S.E. as far as 28° or even to 32° S. latitude. This tack should be kept as far as 32° W., so that on standing to the northward on the starboard tack a vessel may be certain of reaching to windward of the isle of Trinidad. As the vessel proceeds northward the wind will be found more easterly, admitting a slack bowline, or it would be extraordinary if she does not weather Fernando de Noronha, crossing the line between the meridians of 32° and 37° W. From thence the zone of the variables of the equator, generally West of the meridian of 32°, will be crossed, and the starboard tack is kept through the N.E. trade as far as 30° N. latitude. Once beyond the region of the trade winds, the course must be shaped according to the destination, passing northward of the Azores.

After these observations concerning the ports of Brazil, there will be very little difficulty as to the course to be pursued in leaving the southern ports or coming from Cape Horn. The West winds which prevail in this zone will facilitate a vessel's progress to the limits of the S.E. trade.

According to fifteen voyages made from different ports of Europe to Rio Janeiro, the mean duration of the voyage is found to be fifty days. Several voyages have been made from the West coasts of France and England in forty and forty-two days. Packets leaving Rio Janeiro for Englaad generally make the passage in thirty-five days.

Outward Voyage.

The Channel to St. Catherine	77 days.
Straits of Gibraltar to St. Catherine	53 ,,
Havre to Maranhao	43 ,,
Marseilles to Rio Janeiro	65 ,,
Bordeaux to Cape St. Augustine	45 ,,
Bordeaux to Cape St. Antonio	64 ,,
Rio Janeiro to St. Catheri: e	6 ,,
Channel to Montevideo	57 ,,
Channel to Cayenne	31 ,,
Europe to Cape Horn	82 ,,

Homeward Voyage.

Montevideo to Rio Janeiro	11 ,,
Rio Janeiro to the Channel	58 ,,
Ports in the North of Brazil to Europe	33 to 37 ,,
Montevideo to the Channel	83 ,,
Maranhao to the Channel	65 ,,
Cayenne to the Channel	56 ,,
Cape Horn to Rio Janeiro	18 ,,
Cape Horn to Europe	73 ,,

The foregoing will suffice to give a general idea of the voyages from Europe to the coasts of South America. Let us now proceed with those from Europe to that part of the coast of Africa situated North of the equator.

The masters of vessels from the English channel must bear in mind what has been said in regard to the currents of those parts. After having doubled Cape Finisterre, according to the time of year, at a distance of thirty-five or sixty leagues, a vessel should steer between South and S.W., giving the coast of Portugal a wide berth, especially during winter, in order to pass East or West of Madeira or to reach the Canaries, which are always sighted by vessels on their passage to the coast of Africa. These islands may be passed on either side, the channel between them and the African coast presenting no danger which is not apparent. If it is desirable to pass through them, the preferable channel is that between Palma and Hiero on the West and by Gomera on the East.

It is rarely after having passed to the southward by the other channels that calms are not met with, along with a swell which endangers the masts, under the lee of the large islands of the archipelago. This is especially the case with the wind from the North and N.E., which, interrupted by them, does not reunite in a steady course till far to the southward. In November and December it is preferable in bad weather to pass clear away to the westward of these islands in case of meeting the S.E. winds, which are frequent at that time.

If desirous of touching at the Canaries, the best anchorage for a ship is that of Palmas, North of the Grand Canary. The town there offers more resources than those of Santa Cruz, in Tenerife, and the bay is easy to leave under sail in all weathers; which is not the case at Santa Cruz, a harbour generally frequented though very dangerous with a S.E. wind.

On leaving the Canaries, a vessel bound to the coast southward of them will adopt a South, S.W., or S.S.W. course, according as she may have passed outside or through one of the channels of the group. Having passed the parallel of 19° at the southern extremity of the bank of Arguin, she would gradually haul to the eastward till getting thus into the North polar current, in which she would keep her course. St. Louis should be made a little to the northward of its latitude. If bound for Goree, a vessel should pass round Cape Verd. When bound to places South of Goree, such as the Gambia or Sierra Leone, or even to the coast of Liberia, the route as far as Cape Verd will be the same; for a vessel must generally pass it unless leaving the Cape Verd Islands. In all these cases should a vessel not touch at the islands it is best to steer in such a manner as to pass nearer to Cape Verd than to the islands of that name, because the wind is steadier and fresher than on the coast. From Cape Verd the navigation depends on circumstances, being comparatively easy with N.E. winds in the fine season, but difficult with the S.W. winds of winter.

Vessels proceeding to places on the North coast of Guinea or to the isles of the gulf of Biafra or the Gaboon, after having left Cape Verd will make for Cape Palmas, either with the favourable winds from October to May, or with the contrary ones principally during June, July, August, and September, when they blow from S.W., W.S.W., West, and W.N.W., interrupted by calms. At this period it is best to keep 100 leagues from the coast. They will then steer so as to sight the cape or about twenty leagues or more to seaward of it. At this distance they will find the North Guinea current, which is only a continuation of the polar current of North Africa, setting to the East and E.N.E. from 15° or 16° W. longitude. After reaching the parallel of Cape Palmas, they will find, as already stated, the trade wind from S.W. and W.S.W. Winds with a current will then be found favourable for reaching any of the places of North Guinea. But it must be observed that in these routes a vessel should pass further South than 2° N. lat., in order not to get into the equatorial current which sets to the westward. Thus as soon as the parallel of Cape Palmas is reached and the cape sighted by a vessel bound to the Gold Coast or the Ivory or Slave Coast, she should keep in the zone comprised between the coast of Guinea and 2° N. latitude.

The best method for a vessel to navigate this coast is to keep the land in sight, at about the distance of ten or fifteen miles, and to approach it to about the distance of one or two miles when thirty or forty miles West of her destination, taking great care not to run beyond it. In estimating the route it will be very important to consider the velocity of the current—which runs from twenty to twenty-four miles a day—for it is requisite to approach it well to westward of the point to which she may be bound.

If bound for the islands of the gulfs of Biafra or Gaboon, a vessel having doubled Cape Palmas should steer East, keeping between 3° and 2° N. lat. as long as possible, according to the island she is bound for. She should then cross obliquely the zone comprised between 2° N. lat. and the equator, running before the wind for her port, in order to make tho land to the southward of it. The same must be done in going to the islands of the gulf of Biafra. In the vicinity of these islands the current of South Guinea is met with, setting to the N.E., and sometimes N.N.E.; then S.S.W. winds will be found, veering, perhaps, to South as the equator is approached. In sailing from Princes Island to Gaboon the current of South Guinea is crossed, setting N.E., N.W., and sometimes North. It is therefore necessary, in going from Princes Island to Gaboon, generally to make the land to the South of this river in order to counteract the effect of the current.

In the bottom of the gulf of Biafra the currents are variable, although in the latitude of Fernando Po, and between this island and the coast, they generally set to E.N.E. and N.E. If from thence it is desired to proceed to the southward, a vessel should keep at a little distance from the coast of Gaboon, in order to profit by the alternate breezes and to take advantage of the tides.

The current of North Guinea formerly terrified seamen, for they supposed that having once entered the gulf they could not leave it without much difficulty. These fears, as will be seen, were groundless.

Leaving a place eastward of the North coast of Guinea, a vessel should stand well out on the starboard tack till she is clear of the Guinea current and has entered that of the equatorial; and, according to the time of year, she may cross the line to the southward for southerly winds. She may then get on the port tack, so as to reach well to the West of her port of destination, in order to allow for the effect of the current of North Guinea, which will be found in 2° N. latitude; and if she cannot make it so on this tack, she must go about in 2° N. lat. and stand out on the starboard tack again till she has gone far enough West to be sure of reaching the coast to the westward of that port. In a few days, by this method, the port will be gained. Vessels which have endeavoured to get to windward on the coast of North Guinea, are sometimes thirty or forty days in reaching Grand Bassam from Cape Coast, and have been obliged to give up the attempt and stand out to sea.

On leaving Fernando Po, a vessel must make her way along the coast of Gaboon, profiting by the slants of wind and current, and consequently keeping near the coast until she has made southing enough to stand into the equatorial current. Leaving Princes Island she should take the starboard tack, with S.W. winds, and continue on that tack as far as the coast permits; she may then get on the port tack and thus get clear of the gulf of Guinea.

If intending to leave the gulf of Guinea, after reaching the equator she may keep to the southward, profiting as she may by winds from South and S.S.W. to S.S.E., till she reach the meridian of Cape Palmas, and in the case of intending to go to the northward, after reaching about 17° West, she may make for the Atlantic on a course according to her destination. Then, if returning to Europe, it will be best to leave the equator in about 23° W. and make to the northward, and afterwards pursue the same route as that indicated in returning from Brazil to Europe; but if near the equator West and N.W. winds are found, which is often the case during winter from May to September, the ship may then cross it in 17° or 18° W., and pass between the Cape Verd Isles and the coast of Africa. North of the Cape Verd Isles the N.E. trade will be found, which will enable her to proceed on the starboard tack. If returning to any point on the coast of Africa, Sierra Leone, Gambia, Goree, or St. Louis, a northerly course must be taken in 16° or 17° W., and a course made good between the meridians of 22° and 28° W., in order to avoid entering into the polar current of North Africa until the parallel of the Bissagos is reached. This last course will be especially favourable from May to September, which is the winter season. Lastly, a vessel bound to the United States or the Antilles, should proceed North in 28° or 33° W. long.

Favourable Season for leaving the Gulf of Guinea. —The most favourable season for leaving the gulf of Guinea is from May to December. A vessel is then seldom obliged to cross the line; the S.E. winds are generally well established at this period, and reach beyond the equator. But from December to May it is better to cross the equator, and proceed at least in 0° 30′ or 1° S. lat. By following the foregoing directions, a vessel will in a few days be clear of the gulf of Guinea. In order to enter it, a vessel should pass near Cape Palmas, and keep in the North Guinea current, between the coast and 2° or 3° N. lat.

But in order to leave the gulf of Guinea, as a general rule, a vessel should endeavour to reach the equator by the most direct route according to her longitude. From May to December she may keep on the equator, or a little North of it. During the other months it will be better to keep South of 30′ or 1° lat., and to the westward as far as the meridian of 16°, 17°, or 23° W. long., according to the port of destination in the North Atlantic Ocean.

Passages.

North of Europe to Madeira	15 days.
Strait of Gibraltar to Madeira	4 to 5 ,,
North of Europe to the Canaries	16 ,,
Strait of Gibraltar to the Canaries	7 ,,
North of Europe to Cape Verd Islands	20 ,,
North of Europe to Senegal	18 ,,
North of Europe to Goree	20 ,,
North of Europe to Gambia	24 ,,

On the coast of Africa, South of Senegal, the length of the voyage, according to the season, will vary greatly. Thus, in the fine season 28 days are taken to go from Goree to Princes Island, and in the winter generally 36 or 38 days.

Returning.

Princes Island to Goree	38 to 40 days.
Gambia to Goree	3 to 4 ,,
Goree to Senegal	5 to 7 ,,
Senegal to Channel	30 to 40 ,,

There are some instances of this voyage having been made in 24 and 22 days.

Routes from Europe to Ports of Africa South of the Equator.—The routes from Europe to those ports of Africa situated South of the equator, are very different according to the latitude of these ports. They are distinguished as the *Great Route* and the *Little Route*.

The Great Route is that adopted to reach the Cape of Good Hope, and in general all the ports situated South of Cape Negro.

The Little Route is that which ships take to reach ports situated North of Cape Negro. The Great Route is however followed by many vessels bound to these ports. Vessels taking the great route, on leaving Europe will follow the directions given for the routes from Europe to Brazil; they will consequently cross the line between 23° and 28° W. Thence, profiting by the S.E. trade, they will shape their course for the isle of Trinidad. They will pass West of it, and making for the southward will find westerly winds and the counter-current of the South Atlantic. They will then make for the Cape of Good Hope, so as to cross the parallel of 30° S. lat. near 18° W. long. By following these routes vessels have been only 59 days in sailing from the English Channel to Cape Town. A similar route may be adopted when bound to places on the West coast of Africa North of Cape Negro. Thus, after crossing the line between 23° and 28° W. long., a vessel may take the port tack with the S.E. trade and stand on, so that when taking the other tack she may reach the coast to the southward of where she is bound to, and so counteract the effect of the African polar current setting N.W. along the South coast of this continent. But if destined for Benguela, Angola, or even a point North of Cape Negro, the course may be so modified, as above shown, so as to render the passage shorter.

On leaving Europe a vessel should shape her course so as to reach the trade winds as soon as possible, passing either West or East of Madeira and West of the Canaries or in the channels through those islands. Thence she would pass West of the Cape Verd Islands if in winter, that is to say from June to September. During the other months she would pass between those islands and Cape Verd, keeping closer to the cape than to the islands, because near the continent the winds from

N.E. and N.N.W. are fresher and better established in this season. Which ever passage is adopted, after having passed South of Cape Verd she would keep along the African coast, at the distance of sixty or eighty leagues, until the parallel of the Bissagos is passed. From thence she would steer for Cape Palmas, passing it at the distance of twenty leagues, and cross the gulf of Guinea on the starboard tack. This tack will generally enable her to reach Cape Lopez, and often South of the island of Anno Bon. She would then get on the other tack to look for the S.E. winds of the southern hemisphere, and, keeping in the space comprised between the coast and the line passing from the Cape of Good Hope to Cape Palmas, she would again get her starboard tacks on board to fall in with the S.W. winds which prevail there and blow alternately from sea and land from January to September. She would then keep near land to profit by them. The sea breeze lasts during day from ten or eleven o'clock in the morning, blowing from W.S.W. to S.W.; the land breeze lasts the night, from S.E. to South. She would therefore manage her boards in such a manner as to be near the coast for the land wind at night, and to be at a distance from it in the morning for the sea breeze. This navigation is similar to that on the coast of Senegambia in the northern hemisphere; but here the coast is much more extended and the season from January to September is particularly favourable for it. In the rainy season, near Cape Lopez squalls from the westward are sometimes met, but of short duration.

Routes from Europe to the Islands of the South

Atlantic Ocean.—Vessels from Europe to the islands of the South Atlantic Ocean have been sometimes a hundred days on their passage. The following remarks on this subject may prove useful.

To reach Ascension from the channel a vessel should gain the N.E. trade as soon as possible, and pass between the Cape Verd Islands and the continent or else West of the Cape Verd Islands. From thence she would steer so as to double Cape Palmas and make it if she can. As soon as she has lost the N.E. trade she should steer South to cross the zone of the variable winds without passing West of the meridian of 15° or 17° W. In approaching the limit of the S.E. trade winds near Cape Palmas, and even North of this cape, winds from the S.W. will always be found and sometimes from W.S.W. With these she should get on the starboard tack and cross the line in 5° or 6° W. long., or even more to the eastward if she would improve her speed with the currents of the gulf of Guinea. In this case she should make nearly the same course as shown for the islands of the gulf of Biafra, keeping on the parallel of 2° N. in order to reach the gulf of Guinea easily; then cross to the South, reaching Cape Lopez on the starboard tack. As soon as the South and S.E. winds are found she would get on the port tack and soon reach Ascension.

Routes to St. Helena.—There are two different routes from Europe to St. Helena. Considering the position of this island in the S.E. trade it cannot be reached from North without first standing away to the East or West in order to run down on it. The quickness of the passage will depend generally on the time occupied in crossing the zone of the variable winds of the equator. The season will therefore determine which of the routes it will be best to pursue. The western route may always be taken. That of the East is only advisable during the months of November, December, January, February, aad March,—a period when, as above said, the zone of the variable winds of the equator is diminished. The eastern route during the months just mentioned will be the same as that followed in going to Ascension, only the course should be prolonged towards the coast of Africa until the wind fails. The other tack is then adopted and St. Helena is generally reached by this route more quickly than by the westerly one. But when the sun has North declination the eastern route becomes very uncertain and the western is preferable. It may however be taken for granted that a smart sailing ship holding a good wind may adopt the eastern route in all seasons. After having crossed the line between 23° and 28° W. long. a ship adopting the western route will have to get on the port tack. This will take her towards the coast of Brazil, and she must generally tack nearer to it than to St. Helena. On the starboard tack she will then make to the S.E. as far as 23° S. lat., where she will put about again, standing N.E. and North, profiting by wind and current to the East of the isle, having passed South of it.

The currents near St. Helena are not strong, and when the wind is favourable there will be little trouble in reaching the bay, except during the syzigies, when the N W. current prevails.

The average voyage from Europe to the Cape of Good

Hope is about ninety days. Horsburgh, in the *Anna*, made it in sixty-seven days; this is one of the shortest which has been made. A steam-vessel leaving England has reached the Cape of Good Hope in fifty-nine days. The general passage from Europe to St. Helena is sixty days.

Routes from Ascension and St. Helena to the Coast of South Africa.—St. Philip de Benguela being one of the southernmost places reached from Ascension on the coast of Africa South of the equator, we may adopt it, as from it all points to the northward will be easily attained. On leaving Ascension the starboard tack should be adopted, and in order not to fall into the great westerly current a vessel should endeavour not to pass North of the parallel of 4° S. lat., and also not to stand further South when the wind will not permit her to lay about S.E.b.S. true. This however will depend on circumstances. It will be easy to make a short board so as not to pass the above limits; but it will often happen that the passage may be made without tacking at all, because the winds in general along and near the African coast veer to S.W. and sometimes W.S.W. A vessel from St. Helena one can understand has only to lay her head for the point of the coast she is bound for, or something to the South of it, to allow for the polar current of the South Atlantic. And generally, notwithstanding the opinion of several authors who advise that on leaving these islands a vessel should take the port tack to get to the South and West when bound to a place on the coast of Africa as far North as St. Philip, we should take the starboard tack on leaving those islands, and steer for our destination, allowing for the effect of the current, which flows, with a rapidity of fifteen miles in twenty-four hours, to the N.W. and W.N.W.

But leaving Ascension or St. Helena for a more southern part, the Cape of Good Hope for instance, a vessel should adopt the port tack to make southing and stand towards the American coast, profiting then by the remarks made concerning the routes from Europe to the Cape of Good Hope in the southern hemisphere.

Routes from the Coast of Africa to Ascension and St. Helena.—The routes from ports on the coast of Africa to Ascension and St. Helena have been shown in coming from places North of the equator; they are nearly the same as those followed in reaching them from Europe, whether the great or little route is adopted. When a vessel leaves the coast of Africa from anywhere South of these islands the winds and currents are favourable, and she should endeavour to make to windward of it, that is more to the southward than to leeward.

Homeward Routes from the North Coast of Africa to Europe.—In referring to the navigation of the gulf of Guinea the routes from thence to Europe have been alluded to. On reaching 23° W. long., standing to the South of the equator, a vessel should then commence her northing on the starboard tack and cross the zone of the N.E. trade winds. As soon as she has reached the zone of the variable winds she should make progress, passing to the North of the Azores or between them. A vessel starting from a point to the northward of Cape Palmas should get to the westward with S.W. winds which prevail in the vicinity of that cape, passing as quickly as possible through the zone of the variable winds of the equator; she would then take the starboard tack with the N.E. trades and make her northing. A vessel leaving the ports of Senegambia, the Gambia, Goree, or St. Louis with N.E. and N.N.E. winds, during the fine season, would stand out on the starboard tack till she reached the zone of the variable winds. In the homeward routes from points on the West coast of Africa South of the equator that from the Cape of Good Hope to Europe may serve for the rest.

Route from the Cape of Good Hope to Europe.—On coming from the Indian Ocean round the Cape of Good Hope, if in the fine season, a vessel may approach the land without fear and steer North when the cape is passed. But if in the winter season, namely from June to September, before steering North it will be best to get an offing to the West of forty or fifty leagues from the land, in case of meeting with West and N.W. winds, which prevail during this season. After doubling the cape, in the fine season, namely from October to April, a vessel will pass near St. Helena, a short distance either to the East or West of it. From St. Helena she would steer N.W., in order to pass twelve or thirteen miles East or West of Ascension, and thence cross the line between 23° and 28° W. long. The route she would then take has been previously pointed out in returning to Europe. In the case of doubling the Cape of Good Hope between August and September a vessel should keep at a respectful distance from the coast and steer South of the zone of the S.E. trade winds in order to cross the parallel of 20° S. lat. on the meridian of about 18° W. long,

She would then endeavour to cross the line between 26° and 28° W. long. In this season the average of many passages from the Cape of Good Hope to Europe is seventy days; from the Cape to St. Helena generally fifteen days; and from St. Helena to Ascension generally six days.

Routes from the Ports of North America to the Coasts of North Africa.—From the ports of North America to that part of the coast of Africa North of the equator, the course at first is nearly the same as that for returning to Europe; but when a vessel has reached far enough to the eastward to make her port, she would then steer for that part of the coast, crossing obliquely the region of the N.E. trade winds.

Routes from North America to West Africa or South America.—When steering for any port of Africa South of the equator, she would cross the zone of the N.E. trades obliquely, and then the equator between 23° and 28° W. long., and take one of the routes previously indicated, either to the western coast of Africa or the eastern coast of America. In treating of the routes from the coast of Africa to the ports of North America, that from the Cape of Good Hope to these ports need only be pointed out, from which the rest may be easily deduced.

Leaving the Cape between October and April the prevailing winds will be found from S.E., and the course will be the same as that previously shown for returning to Europe until the equator is crossed in 28° W. From thence a vessel would pass West of the shoal called Penedo de St. Pedro and proceed with the trade winds from East or E.N.E. in order to pass at a good distance to windward of the Lesser Antilles. This course, as may have been seen, presents no difficulty.

A ship rounding the Cape of Good Hope between the months of March and September, should avoid the coast on account of the N.W. winds, which blow with violence during the winter, and keep South of the zone of the trade winds in order to reach the parallel of 20° S. lat. in the meridian of 18° W. She would then steer northward and cross the line in 33° W. long. During this season it is preferable to cross the equator on this meridian rather than in a more eastern one. It is also better to pass East of the Bermudas if she should be bound to a port of Nova Scotia, instead of West of them, because at this period easterly winds are often found in those parts. The rule generally adopted is to pass East of the Bermudas from the middle of March till October in going to any port on the North coast of America situated North of New York.

From the different routes now pointed out it will be easy to design any that may be required from one point to another of the Atlantic Ocean.

A CONCLUDING GENERAL VIEW ON MAKING PASSAGES IN THE ATLANTIC.

THE principles to be observed in making a passage between two places, whether under sail only or with the assistance of steam, are:—1.—Never to hug the wind when it is foul, but to let the ship go at least a point free through the water; and, 2.—To profit as much as possible by the well known prevailing currents of the several seas that are happily so distributed throughout them as to be favourable for the navigator in different latitudes when he is desirous of reaching either shore from any place, whether this be on the same or on the opposite coast; and 3.—To lay the ship's head on that tack, with a foul wind, that will enable her to look best up for her port.

All currents are more or less influenced by the wind. Before therefore commencing with the mode of making passages, it may be as well to take a rapid view of the prevailing winds which are to regulate the seaman's course across those seas.

Within the tropics the prevailing wind is generally easterly; to the northward of the equator as well as to the southward of it, it draws slightly towards either pole according as the sun retires from it. These winds, called the Trade winds, prevail throughout the equatorial regions of the Atlantic and Pacific Oceans; but in the Indian Ocean they are modified by the great African continent,—by which their character is totally changed; and according to the position of the sun (as they generally blow towards it) they become monsoons or periodical winds, but still subject to certain modifications.

The surface of the globe between the tropics and either pole is the region of the variable winds; generally assuming a contrary direction to the trade winds, to contribute, perhaps, with other phenomena to preserve the atmospheric equilibrium or counterpoise that is found in all the operations of nature. But the whole subject of atmospheric changes, like that of natural history generally, is replete with those beneficial arrangements of an all-wise Creator in anticipation of the wants of his creature man to reach readily the different parts of the globe which he inhabits by the aid of navigation. So that he may turn to his account the winds and the currents which again and again change their directions with the seasons of the year. All is ceaseless change, perpetually working to his good,

Such is the reflection resulting from a general view of the winds; their several peculiar modifications will appear as we proceed and have to allude to them in showing the manner of making the several passages on which we propose to treat.

A ship leaving the British Channel for any port to the southward should first gain a good offing, and then shape her course according to the wind, modifying it according to that which may be expected as she proceeds. In crossing the bay of Biscay it was customary in former days to make an allowance for what was considered the indraught to the eastward;—but modern navigators have shown that the reason of ships finding themselves to the eastward of their reckoning in proceeding southward, has arisen more probably from the deviation of the compass caused by local attraction. Much however will depend on the direction of the wind and how it has previously been. If it has been to the northward of West, the great body of waters accumulated in the bay may be looked for as finding an escape to the westward along the North coast of Spain, and thence off Cape Finisterre to the S.W., eventually joining the Portugal current running to the southward. But even this is itself influenced by the wind which has produced a northerly current there as strong as that to the southward. Again, if the wind be to the southward of West the accumulated waters of the bay may be expected to run to the northward, and thence occasion an easterly set along the North coast of Spain.

Thus a vessel bound to a port on the Spanish coast in the bay will endeavour to make the land to the eastward or westward of it according as she may have had the wind, so as to make the land to windward of it or not (by the previous winds experienced) of that port. But for a vessel bound to the southward of Cape Finisterre a S.W. course is recommended in order to shut out the British Channel, and to reach fine weather as soon as possible. And having gained the latitude of 45°, should the vessel be bound to Oporto or Lisbon it might be well to make the land about Cape Toriñana, as by route A. and close with it if going to the former port. If to the latter, the vessel will make the best progress she can to the southward with a good offing of 30 to 60 miles from the coast, and make the rock of Lisbon from whence she will gain her port. If bound to Gibraltar or up the strait she will still be in a good position for that destination. The passage from the Channel to Lisbon or the Straits will vary from four or five days to a week or more, depending much on the direction of the wind.

In the case, however, of a vessel bound southward, as soon as she has reached the parallel of 45° in about 12° West, she would shape her course direct for Madeira, as by route B, or if not desiring to touch there, for the Canary Islands. In case of going to Madeira, Porto Santo should be made, and from thence the East end of the island will be passed and Funchal Roads will be gained. In the case of a vessel stopping at the Canaries, she will probably round Point Anaga, and anchor off Santa Cruz. The anchorage off Palma is preferred to that of Santa Cruz, as having, it is said, more resources and being easy to leave, while the latter anchorage is much exposed to S.E. winds, and is not so easily left as Palma. But if it be not desired by a vessel to touch at any place on her way southward she may continue her course, passing to the East or West of these islands and reach the trade wind as soon as she can. She will generally be more certain of a breeze by passing to the westward of the islands. The passage to Madeira from the Channel is considered to occupy about a week.

Continuing her progress to the southward on the eastern side of the Atlantic, should a vessel be going to the coast, the Cape Verd Islands may be her next place of call, for which a S.W. course will be proper. In this route there is nothing to observe upon further than that she has more chance of a breeze by keeping her distance from the coast. In June the passage inside the Cape Verd Islands may be taken, on the meridian of 20° W., with advantage.

If bound to the Gambia or Sierra Leone, whether from the Cape Verds or the Canary Group, the ship would endeavour to make Cape Verd as by the inner route C, from whence she would have the current and wind in her favour, and make the land to windward of her port. From either of these places, if required, the route to any part of the coast is easy by keeping within the limits of the easterly current, the shortest distance of which from the coast is about seventy miles off Cape Palmas: and in this manner, by keeping in this Guinea Current along the shore, as in route D, or the westerly current in the offing outside of it, passages are made to and from ports as far as Fernando Po and Princes and St. Thomas Islands. But to the southward of those ports to reach places on the coast, as the prevailing wind all the year is between S.W. and S.S.E., then a ship must get to the southward of her destined port with a good offing, and allow for a northerly current also while she is standing in for the coast on the starboard tack, taking care always to make it to the southward or windward of the port to which she is proceeding. And this observation refers to the whole extent of the coast to the southward, but as the cape is approached N.W. or S.W. and even S.E., winds will be mostly found, the latter especially being very heavy.

A vessel bound to Ascension from the Canary Islands from November to February inclusive, would do well to make good a course so that she would pass about midway between the Cape Verd Islands and the coast, as in the inner route C. Passing these islands she might with advantage avail herself of the Guinea current to help her down to the N.E. of that island, along route D, and would thence stand across the equatorial current to the southward and make Ascension with the S.E. trade. If bound to St. Helena from thence, she must stand on across the trade and get into the zone of the variable westerly winds to the southward of its latitude, as shown by route E, and navigate so as to reach to the S.E. of the island, from whence she would run down to it with the prevailing S.E. winds. But in the other months a ship bound to Ascension or St. Helena, might do better by taking the route C across the equator in about 20°, and standing to the southward to about 15° S., so as to be able to lay some points to windward of it with the S.E. trade, making it with a fair wind from the southward. And in the same manner St. Helena would be gained, as shown from the route C or that of E from Ascension.

A vessel bound to the Cape should make the best of her way to the southward by routes B and C, which will take her as far as the Cape Verd Islands. From them she would cross the equator in 25° to 30° West, standing across the S.E. trade, and making the best of her way to the zone of the variables, crossing the parallel of 30° S. in about 15° to 20° West,(see route F,) where she will meet westerly winds, with which she will soon reach the Cape. This passage has been made to the eastward, but the S.W. winds near the coast met with as she gets to the southward, are so heavy as well as unfavourable that she is unable to make so good a passage as by the route here pointed out.

Returning home from the Cape a vessel may shape her course by the route L, which will take her by St. Helena and Ascension, from whence she may cross the equator in about 20° W., and follow that route to the northward as the wind may permit, observing the maxim with which we set out, of letting the ship make her way a point free from the wind.

Passages to the western ports of thc Atlantic from the Channel vary as to the mode of making them according to their position. Thus to ports of Nova Scotia, Newfoundland, or the St. Lawrence, the nearest distance is by the northern route, G, making good as far as 53° N. latitude, and then, as the coast is approached, there is the chance of making it with N.W. winds; while by standing to the southward at first, the whole force of the Gulf Stream current has to be met. But in shaping her course with a contrary wind, a ship should adopt that tack on which she will make most westing.

In crossing the Atlantic for any of the western ports, there are few months of the year in which a vessel may not expect to meet with ice. Happily the sea is less encumbered with it in the dark nights of winter than in summer, when there is so much more daylight that it is less difficult to avoid. Drifting down from the shores of Greenland and Labrador, it is most commonly found in the Atlantic in the earliest part of our summer, and lingering far to the eastward to the last days of our autumnal quarter. But it is most abundant near the banks of Newfoundland, in some parts of which probably much about the Virgin Rocks. Between these rocks and Newfoundland the ice is most abundant, completely lining the coast in the severity of winter, and followed up in the spring by the drift ice on which the seal fishery is followed. Eastward of the banks the commonly assigned limits to their dangers are 40° E. long. and 40°. N. lat. see * on the chart. Solitary cases there have yet been of masses having been met with far to the eastward of this position, and therefore having commenced the passage across the Atlantic whether by the northern or southern route, the most vigilant look out is required to avoid running into them. It has been proposed by the American officer Lieut. Maury, for steam-vessels to make the passage by two distinct routes, one to be adopted by vessels from Europe to America, and the other by those from America to Europe. The chances of vessels running foul of each other would be lessened by this method, and it was proposed that the northern route should be adopted by vessels from Europe to America, and the other, about sixty miles to the southward, should be followed by those from America to Europe; a judicious arrangement, by which those vessels going West will be less within the effects of the Gulf Stream, while those from the westward will be always in a position to profit most by it.

A vessel bound to Halifax, following up the route G, will endeavour to make the land to the southward of her port, where a remarkable distinguishing feature, alluded to in the directions, enables her to recognize it, and thence how to approach it; but in making for New York or any port of the States, she will ascertain by her thermometer when she is crossing the Gulf Stream. Nor must her commander be surprised at passing two distinct currents of warm water, which, having done a counter-current setting to the southward, will be found along the American coast. But in making the passage it is recommended not to pass to the northward of Sable Island, in order to avoid being set down by the tides into the bay formed by the two extremes of that island, slightly curving as they do to the northward. The passage to Halifax may occupy thirty days, while that to New York takes thirty-six days to perform; the return passages to the Channel generally requiring less.

A vessel bound to any of the West India Islands or ports of the continent within them, should make her way to the S.W., so as to pick up the trades near Madeira. She may probably have to touch at that island. But she will find the trade wind earlier or later when in its neighbourhood according as the sun has southern or northern declination. And she will make the best of her way to the S.W., shaping her course according to the southern or northern position of the island to which she is bound, as shown by the routes H and J. But if between May and December, she should keep as much as she can along the parallel of 19° N., and if between December and June, she should keep further South, taking care to make due allowance for the current, which will always place her ahead of her position by the reckoning, and will at least amount to about twelve miles per day. If bound to any of the large islands or the coast of Mexico, she would enter the Carribbean Sea between Martinique and Antigua, by route H, or if to

Cartagena or Porto Bello, she would pass on either side of St. Lucia, and fall into route K. Again, if bound to either of the ports of Guayana or Venezuela, she would shape her course by route K, so as to make the land well to the eastward of her port; taking care that she is not drifted past it by the current, which will be found to acquire strength as she nears the coast. But to do this a vessel should stand South across the N.E. trade and the variables, so as to reach the S.E. trade and edge away to the part of the coast according to the position of her port.

A vessel homeward bound from the West Indies would adopt the eastern route O, or the western ones M or N, by the Florida Stream, according as she is to windward or to leeward. Thus from St. Thomas or any of the Carribbee Islands, she would stand to the northward by adopting nearly the route N. If from Jamaica, she may either pass between St. Domingo and Cuba, along the Old Bahama Channel North of Cuba, or run to leeward round its western end and pass the Havana up the Florida Channel by the route M. This is obviously the homeward route for vessels from the gulf of Mexico, while those from Guayana and Venezuela may adopt the route N, those from the latter gaining it by passing through the Mona Passage between St. Domingo and Porto Rico,—or they may take that between Cuba and St. Domingo, and thence join the route M.

It has been observed by an officer of considerable experience in West India cruizing that the probable best homeward route from Jamaica is by the windward passage between St. Domingo and Cuba, thence by the

Crooked Island Channel to the N.E., thus joining the route N. This is best adapted however to smart sailing ships; but that most generally adopted, especially when so far to leeward as Jamaica, is to run westward round the West end of Cuba and so up to the northward with the Gulf Stream, by the route M.

A vessel bound to any port of the northern coast of Brazil, will adopt the route K or F as far as the equator, from whence she would shape her course so as to be on the coast well to the eastward of the port to which she is bound, and be careful not to be set to the westward of it by the current. A vessel bound to Pernambuco or Bahia will equally observe the route F as far as the equator, always making the land to the northward of either of those ports, and allowing for the Brazil current, which runs along the coast to the southward. From October to March, when the sun is to the southward, the wind is directly on shore, and the current sets to the southward; but from March to October the current in shore sets northerly.

A vessel bound to Rio or Buenos Ayres, will continue on her route F, and approaching the coast gradually, will make it about Cape Frio, by which she will have all the advantage of the Brazil current. On leaving any of these ports for Europe, a vessel has to gain if she can the homeward route L, or to cross the equator as near it as she can, and gain it as she can, although she may not be able to do so even as far North as 30° lat. But to reach a port on the opposite coast of Africa, it is obvious that her proceedings must depend on the latitude of the port which she is leaving as well as that to which she would proceed. It is clear that on any part of the Brazil coast she is dead to leeward of her destination, and she has to cross the Atlantic to the southward or the northward as most favourable to her.

The passage from Rio to the Cape is sufficiently favourable. A vessel will stand to the S.E., as on route P, and soon find variable winds, and as she makes southing along with her easting will pick up westerly winds, which will speedily carry her to the Cape on this route; and vessels from Bahia and Pernambuco for the Cape would follow a similar course, first not being afraid of making southing. In fact, having gained the region of southerly and S.S.W. winds between 20° and 30° S., a vessel may run in upon any part of the African coast that she requires to make. But if she be to the northward of Pernambuco, or even at Cape St. Roque, she must make the best of her way across the trade to the northward, making a detour, and reaching as high perhaps as the latitude of Madeira, till she finds the variables, in which she may make her easting, and thence run down and take any port she may require on the coast by the route C, already mentioned. Vessels from the African coast coming to the coast of Brazil, and desiring to return, should keep to the southward of Cape St. Roque, as the return passage from thence is incomparably shorter to the southward of the S.E. trade than to the northward of the N.E. trade, especially when the sun is to the southward of the equator.

It is to be observed that the several routes delineated on the chart and here referred to, are those expressing nearly the tracks which ships are recommended to follow. But as their ability to do this will much depend on the wind that they may severally meet with, the different routes must be considered as merely meant for their guidance, to be followed as nearly as the winds will admit.

INDEX.

LONDON:

WALTER SPIERS, PRINTER, GREAT PRESCOT STREET.

ATLANTIC OCEAN. PREVAILING WINDS AND TRACKS.

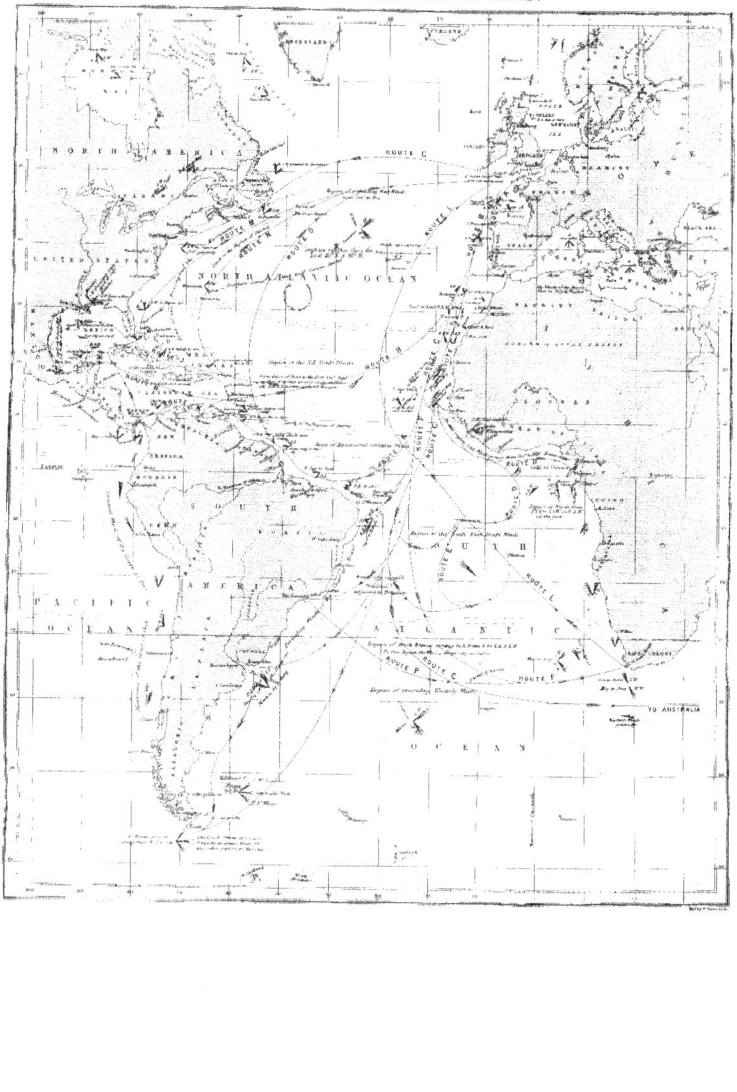

CHART OF THE ORDINARY CURRENTS OF THE ATLANTIC.

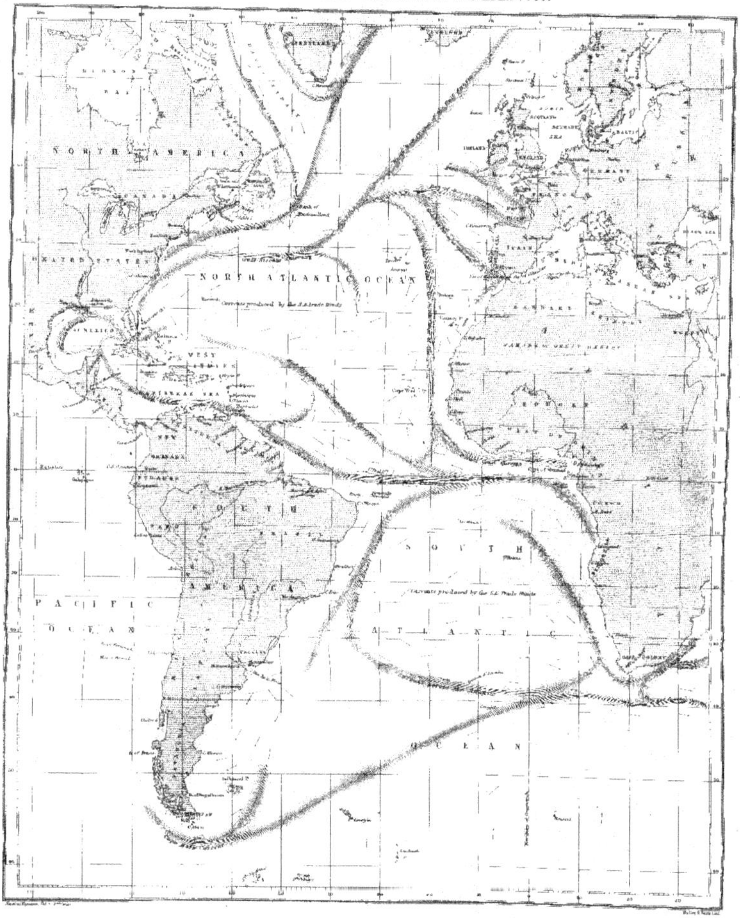

THE

INDIAN OCEAN.

INTRODUCTION.

The welcome reception which the first edition of this volume has met with from those hands for which it was intended, even as it came from the pages of the Nautical Magazine, is a satisfactory proof that its useful nature is duly appreciated.

Another edition, with various improvements, is here completed, and although the three parts of which it is composed contain the body of our information on winds, currents, and the mode of navigating the seas on which they treat, they must still be considered as far from perfect. Any information therefore with which the Editor may be favoured relating to either of them throughout those extensive seas, will be thankfully appreciated.

CONTENTS.

Part I.—Winds.

Part II.—Currents.

Part III.—Navigation.

INDEX.

THE

INDIAN OCEAN.

WINDS.

CHAPTER I.

A GENERAL INTRODUCTORY VIEW OF THE MONSOONS AND TRADE WINDS OF THE INDIAN OCEAN.

The Indian Ocean is bounded on the North by the South coasts of Asia; on the East it is separated from the Pacific by the islands of the Indian Archipelago; Africa separates it on the West from the Atlantic Ocean, and its Southern boundary is generally considered to be a line drawn from the Cape of Good Hope to Bass Strait. The peculiar form of the land surrounding this extensive basin prevents the currents of sea and air being so regular as they are in the Atlantic Ocean.

The S.E. Trade Wind.—In the Northern and even in part of the Southern hemisphere of the Indian Ocean, Trade winds are not met with. In this ocean they are only found between the parallels of 10° and 28° S. from the West coast of Australia to within a few degrees East of Madagascar. The southern limit of the S.E. Trade is influenced by temperature: it shifts to the North or South two or three degrees according as the sun has North or South declination.

It is the same also with respect to the equatorial limit, which, under the same conditions, approaches the equator more or less, and sometimes even reaches it.

If the sun be in the northern hemisphere, at its greatest distance from the equator, the Trade winds have a tendency to come more from South, varying from S.E. to S.S.E.: when the sun is in the southern hemisphere, the winds take a more easterly direction, varying from E.S.E. to East, and sometimes to E.N.E.

Variable Winds.—From the parallel of 10° S. lat. northward to the coast of India, periodical winds are met with in the Indian Ocean. The Trade winds of this sea are only regular at a considerable distance from the land, because the continents and large islands obstruct the regular course of these atmospheric currents.

Variable Winds between 28° *and* 60° *and* 70° *South Latitude.*—South of 28° S. lat. is the zone of the variable winds. In this region, between 28° and 60° S. lat., there is a constant strife between the inferior currents of air flowing from the pole towards the equator and those which are returning from the equator towards the pole. The latter currents, occasioned by the Trade winds, have received, as already observed, the name of tropical currents. This opposition, in which the tropical wind generally prevails, produces the variable wind of this zone.

In the southern hemisphere of the Indian Ocean, southward of the Trade wind, as in the same hemisphere of the Atlantic, the prevailing wind is N.W. during the greater part of the year.

It may be stated generally, that in the zone comprised between the parallels of 60° and 70° S., the prevailing winds during summer are mostly from South, and in winter from West.

Periodical Winds or Monsoons of the Indian Ocean.—The Trade winds which blow fresh in the Atlantic South of the equator, and sometimes even South of the parallel of 6° N. lat., are, as previously stated, interrupted by the peculiar formation of the land constituting the basin of the Indian Ocean on the N.E., N.W., and East. The influence of these lands combined with the constant difference between the temperature of the continents and that of the adjoining sea, occasions the phenomena known by the name of the *Monsoons.*

Monsoons are found in several quarters of the globe; but in no sea do they blow with such regularity, or are they so well established, as in the Indian Ocean, especially in that part North of the equator. There they extend from the coast of Africa beyond the China Sea, and even penetrate the Pacific Ocean; although in the northern part of the China Sea and further East they are less regular than in the Indian Ocean.

The monsoons generally blow towards the continents during the summer, and in an opposite direction in the winter. Thus, the S.W. monsoon, which blows in the North part of the Indian Ocean from April to October, corresponds with the season when the sun, having attained a high North declination, has warmed the surface of the countries of India; while during the winter in this hemisphere, when the sun has a South declination, the N.E. monsoon prevails.

Duration of the Monsoons.—The monsoons blow in one direction during one half of the year, or rather from the middle of April to the middle of September; and from the middle of October to the middle of March in the opposite direction.

S.W. and N.E. Monsoons.—In the North Indian Ocean the S.W. monsoon begins in the middle of April and terminates in the middle of September. The N.E. monsoon follows, and lasts from the middle of October to the middle of March. The S.W. monsoon brings rain and bad weather. The wind is stronger during this monsoon than in that from N.E., when the weather is generally fine.

Zone of the S.W. and N.E. Monsoons.—The zone comprising the S.W. and N.E. monsoons extends from the equator to the tropic of Cancer; from the East coast of Africa to the coasts of India, China, and the Philippine Islands; and their influence is often found in the Pacific Ocean as far as the Marianne Islands; that is, 145° E. long. In the North they sometimes extend as far as the Japan Isles. The limits of the monsoons are not everywhere the same, and their change does not always take place at the same period. It must be observed that in the gulf of Bengal the wind varies more in strength and direction than in the Indian Ocean, where it rarely fails at the expected times, which are generally at the changes of the monsoons.

It is also observed, that in that part of the Indian Ocean where the monsoons are found, the wind has more Southing near the coast of Africa, and more Westing on the coast of India, during the S.W. monsoon.

South of the equator the S.E. monsoon commences in the middle of April and terminates in the middle of September. The N.W. monsoon, varying to W.S.W., follows it, and commencing in the middle of October terminates in the middle of March. This last monsoon is the season of squally bad weather.

In the Indian Ocean the S.E. and N.W. monsoons blow in a zone comprised between the equator and the parallels of 8° or 9° S. lat.; but on the coast of Australia, and in the West part of the Pacific Ocean, this zone extends as far as 12° and 14° S. latitude. It begins in the West near the coast of Africa, and terminates in the East on the meridian of the islands of Pomoutou and Nouka-Hiva, situated in the Pacific Ocean in about 138° W. longitude.

The N.W. monsoon rarely blows either strong or regularly, excepts in the months of December and January, at which time it occupies in the Indian Ocean a zone comprised between 10° or 12° S. lat. and 2° or 3° N. lat. This monsoon is also subject to many irregularities. In proportion as we approach Sumatra its northern limit recedes more South from the equator, and leaves a space between itself and this great circle, to be occupied by calms and variables.

The S.E. monsoon, which is the period of the fine season South of the equator, may be considered as an extension of the S.E. Trade winds, which then blow as far as the equator, when the sun is near the tropic of Cancer.

The S.E. and N.W. monsoons are found with the greatest force and regularity in the Java Sea, the Sea of Timor, and among the Moluccas, and especially near New Guinea. These monsoons are felt on the North coast of Australia, between Melville Island and Cape York, as well as in Torres Strait, where the N.W. monsoon begins at the end of October. In these seas they are often called the West and East monsoons.

Change of Monsoons.—The change of monsoons takes place between the latter end of March and September, and the beginning of April and October. In some latitudes it takes place a week or two sooner than in others. The change is gradual, and is generally accompanied by squally and tempestuous weather. However, South of the equator, if the weather is sometimes bad at the change of the monsoon, the wind is never strong. When the monsoon has nearly terminated, the highest clouds are observed taking an opposite direction to that of the wind, and this happens some weeks before the change is perceptible. The monsoons penetrate far into the interior of the continents bounding the Indian Ocean on the North, and their direction is then influenced by the form of the coasts and islands, the chains of mountains, or other causes.

Seasons in India.—The direction of the monsoons in India regulates the dry and rainy seasons. The rainy season of the West coasts corresponds with the S.W. monsoon, and that of the East coasts with the N.E. monsoon; in other words, the winds from the sea in general produce rain, while those from the land bring fine weather.

Monsoons on the Coast of Madagascar, East Coast.—It has been already observed that the S.E. Trade wind does not extend as far as the coast of Madagascar. On the East coast of this island, and in the channel separating it from the coast of Africa, we find the monsoons regularly established.

N.E. Monsoons varying to N.N.E.—On the East coast of Madagascar the N.E. monsoon, varying to N.N.E., blows from November to April, and the S.E. monsoon from April to November, the last being the time of the fine or dry season. However, in the S.E. part of Madagascar, we generally find N.E. winds, which, veering round the coast, blow as far as Cape St. Mary, and West of this cape and off the S.W. of Madagascar, blow from the S.E. to South during the whole year; but do not reach far to the North along the West coast of this island. Land winds are not known at Madagascar; on all the coasts of this island the winds abovementioned prevail throughout the year, and those to which we shall presently allude in the Mozambique Channel. These winds commence about eight or nine o'clock in the morning, freshen up towards noon, when they blow very strong, and continue so till three o'clock, then gradually abating towards sunset.

Monsoons in the Mozambique Channel.—In the Mozambique Channel, and chiefly between the parallel of the bay of Sofala and the equator, the S.W. monsoon begins in April and terminates in November.

S.W. and S.S.W. Monsoon.—This monsoon forms

the period of the fine season. At the entrance of this channel we find the wind generally from South to S.S.W., veering westward as we go northward; and they unite at last with the W.S.W. and S.W. winds, blowing at the same time North of the equator. In the North part of this channel, during the S.W. monsoon, the winds vary from S.W. to S.E. and E.S.E., and near the southern end of Madagascar they are often very fresh, blowing from S.E. and East. Near the coast of Africa the land breezes are frequent. In the middle of the channel the winds are more regular and follow the direction of the channel.

N.E. and N.W. Monsoon.—During the remaining portion of the year, from November to April, we find in this part of the sea, as far as the northern point of Madagascar, strong N.E. winds, accompanied by rain and stormy weather; while between this island and the extremity of Mozambique, and as far South as the tropic of Capricorn, we find during this monsoon winds varying between N.E. and N.W., with bad weather; forming, in the Mozambique Channel, the period of the rainy season. When the N.E. monsoon, varying to N.W. in the Mozambique Channel, meets with fresh South and S.E. winds at this period near the South part of Madagascar, and even veer northward in part of this channel, the meeting occasions violent and sometimes rotatory storms. The meeting of the North winds with those from S.E., South, and S.W., which blow constantly South of the tropic of Capricorn, sometimes produces the same result.

Land and Sea Breezes.—Alternate land and sea

breezes are found on some of the coasts of the Indian Ocean as in other seas. These breezes generally extend only a short distance from the shores. The sea breeze lasts during the day, and the land breeze during the night. They occur principally about the change of the monsoons; but when they become regular, these alternate winds almost entirely cease.

Nevertheless, on the coast of Malabar, from September to April, the period of the fine season, the East wind blows from the land towards midnight, and lasts until ten or eleven o'clock in the morning, extending sixty miles out to sea.

On the whole coast of India the land wind, where it is found, rises towards midnight, and the sea breeze blows from noon till night. When it continues after sunset it is generally moderate and often slight.

CHAPTER II.

PARTICULARS OF THE MONSOONS, LAND AND SEA BREEZES OF THE AFRICAN COAST AND MOZAMBIQUE CHANNEL FROM THE CAPE OF GOOD HOPE TO THE RED SEA.

Winds of the African Coast from the Cape of Good Hope to Cape Corrientes.—From October to April strong South and S.E. winds prevail along the coast of Africa from the Cape of Good Hope to Cape Corrientes, and sometimes reach as far as the equator. From May to October, on the same coast, the prevailing winds are West and N.W. for a distance of thirty leagues at sea. These winds are only found between the Cape of Good Hope and Cape Corrientes. In May they become light and interrupted by calms. The N.W. winds are the strongest, and bring rain, and are especially violent in June, July, and August.

Near the bank of Agulhas, as well as in the greater part of the southern hemisphere, the mercury of the barometer rises with southerly winds and falls with those from the northward. From the bank of Agulhas as far as the meridian of the southern end of Madagascar, we find, during the winter, violent winds, accompanied with thunder, lightning, and a great deal of rain.

Winds on the Coast between Cape Corrientes and the Equator.—On the coast between Cape Corrientes and the equator the monsoons are found abovementioned, as those of the Mozambique Channel, and also alternate land and sea breezes. The seasons are very uncertain, the winds variable, and tempestuous weather frequent at all times of the year, especially during the N.E. monsoon varying to N.W., in November to April. The fine season lasts from April to October, the winds from South to S.S.E. and S.S.W. prevailing.

Winds on the Coast between the Equator and Cape Guardafui.—On the coast between the equator and Cape Guardafui the S.W. monsoon from W.N.W., varying to S.W., according to the portion of the coast, sets in about the beginning of March. The N.E. monsoon succeeds after six months (from October to March). The monsoons are established near the coasts much sooner than they are felt out at sea. They also begin near the line before they are felt at Cape Guardafui. The N.E. monsoon is seldom strong on this coast.

From October to January the wind is variable, and frequently blows in succession from all points of the compass. The North is the prevailing wind, brings rain, and sometimes blows strong.

From January to May it is fresh from N.E. and N.N.E., with fine weather. Towards the end of March S.W. winds are found near the coast, which gradually extend out to sea. During this latter month tornadoes are frequently found near Cape Guardafui.

From May to October the wind is from South, and during July, August, and September there are calms near the coast, while at the distance of 100 leagues a fresh wind is blowing from South. During these months near Cape Guardafui strong winds are found, which do not extend more than thirty miles out at sea; at which distance the weather is fine with a smooth sea.

Alternate Land and Sea Breezes.—Along the East coastof Africa, during the month of February and the first part of March, we find alternate land and sea breezes, ceasing, however, during the S.W. monsoon. When this monsoon loses its force they re-appear, during October, a period when, however, they are not so fresh or so regular as at the end of the preceding monsoon.

Gulf of Aden.—In the gulf of Aden, comprised between the meridian of Cape Guardafui and that of Babel Mandeb, during the months of January, February, and March, winds from East and E.N.E. are found successively from the promontory of Aden on approaching the strait. The weather during these months is fine and generally clear. At this period the trade is carried on by vessels of from fifty to three hundred tons.

In April and May the wind is generally light, varying from E.N.E. to S.E. and South. Near the coast the weather is fine; sometimes, however, the sky is obscured, the horizon misty, and land breezes are felt from four o'clock in the morning until eight. Westerly winds are not frequent, but sometimes strong. April brings rain, but rarely lasting more than three days. During the night the dew is very heavy.

In June the weather is very uncertain, the wind variable, the sky sometimes clear and sometimes very cloudy. In the morning there is a calm or light winds, sometimes increasing towards noon and even blowing fresh from the South and causing a great swell on the Arabian coast. About the middle of this month, between Burnt Isle or Mait and the strait, are strong westerly winds, which enable vessels leaving the Red Sea on their voyage to India to reach the monsoon. In July and August nearly the same weather continues. It is almost always cloudy and fine days are very rare; sometimes for two or three consecutive days there is a thick fog.

According to the mean result of six years' observations, in sixty-two days there were twenty-four days of South wind, fresh or moderate during the day and light at night, but with a heavy swell on the coast of Arabia. During the remaining thirty-eight days the wind was from West and S.W., and in the two months alluded to there were sometimes passing showers in the strait and the dew abundant at night. Sometimes, towards evening, when the South wind is no longer felt, a violent storm comes off the land, announced by a thick cloud of sand or very fine dust, giving vessels time to prepare for it.

In the month of September the West winds cease and are replaced by alternate land and sea breezes. During the night there is often a dead calm.

In the beginning of October the weather is nearly the same, with the exception of a little rain. In the beginning of November the N.E. monsoon, which reaches Makallah about the 5th of November, gradually increases, and blows fresh about the time of the syzigies. During these months the winds generally vary from

East to E.N.E., and for a period of four years it is stated that from the 27th of December to the 3rd of January the weather was constantly threatening, with a storm, accompanied by a great deal of rain.

South of the Red Sea the S.W. monsoon of the Indian Ocean becomes more southerly, but is obstructed by the mountains of the Arabian coast. Beyond this strait the monsoon blows from the West, and rarely passes beyond Aden. At Cape Guardafui it blows freshly from N.N.E. near the coast, and the breeze then enters the gulf of Aden and reaches as far as Ras-Rhemat, a cape situated S.W. of Makallah. On this parallel vessels from the Red Sea, going eastward, generally take the monsoon.

CHAPTER III.

MONSOONS OF THE RED SEA, PERSIAN GULF, AND ARABIAN SEA, AS FAR AS CAPE COMORIN, INCLUDING, WITH THE MALABAR COAST, THE LACCADIVE AND MALDIVE ISLANDS.

FROM the beginning of October to April, called there the winter season, the wind is constantly from South between the straits of Babel-Mandeb and Jibbel-Teer, situated in lat. 15° 30′; at the time of the syzigies it is interrupted for two or three days by North winds. There are, however, two whole months without any change in this monsoon.

From Jibbel-Teer to the parallel of 19° or 20° N. lat., the winds are variable during this season, and come as often from North as from South.

From the parallel of 21° to 27° N. lat. the North wind prevails from October to April. However, it is rare that a fortnight passes without a day or two of South wind, especially from the end of November to the beginning of March.

From the parallel of 27° N. lat. as far as Suez, the wind is generally from North, and rarely becomes southerly except during the months of December, January, and February. The N.W. wind is seldom strong for more than twelve or fifteen hours, and it is more violent in winter than in summer.

From June to September northerly winds blow without interruption, and with more or less force over the whole extent of the Red Sea from Suez to the straits of Babel-Mandeb. In August and September it varies towards the land. During these months a merchant vessel can gain to windward thirty-five miles a day in working up between Mocha and Suez.

In the Red Sea during June, July, and August the wind is stronger at night than in the day.

In that portion of the sea called the Sea of Suez a misty horizon is generally the forerunner of wind, which however does not always follow. The same remark applies to the little fleecy clouds observed over the mountains of Toor or Sinai seen at the entrance of the strait of Jubal.

During the winter months the northerly winds bring dry weather throughout the whole extent of the Red Sea, and the southerly winds wet weather. A change of wind is thus announced previous to any change being observed. During the summer months the atmosphere is generally moist over the whole extent of the Red Sea, with a clear sky over head in the zenith. The wind rarely blows in sudden gusts in this sea; sometimes, however, it is strong from the North.

Each monsoon takes a month before it becomes completely established. At Mocha, from September to May, the wind varies from E.S.E. to S.E.b.E., and from the beginning of April to the end of August it frequently blows from S.S.W. to S.W. In November and December, along the coast between Mocha and Aden, the breeze is very strong; in June and July it is variable and moderate.

The rainy season in the Red Sea lasts from April to September. When the N.W. monsoon penetrates into the Red Sea, it assumes the direction from S.E. And enclosed as it is by the high lands of Africa, in a narrow strait, it gains considerable force, inclining slightly towards the coast of Arabia. Indeed it is probably stronger on this coast than on that of Abyssinia, even in the lower part of the Red Sea. It begins to diminish in force towards the parallel of 14° N. lat., and in proportion as it reaches the widest part of this sea, it gradually becomes a light breeze, turning West of the shores and islands of the coast of Africa, and meeting the northerly winds of these parts.

South Coast of Arabia.—On the South coast of Arabia beyond the gulf of Aden, that is, West of the meridian of Cape Guardafui, as far as Ras-el-Gat, in December, January, February, and the first half of March, the N.E. monsoon prevails along the coast, with a direction depending on the formation of the shore. At sea, it varies from N.E. to E.b.S., with fine clear weather, without any squalls or rain.

Koorya-Moorya Bay.—The same winds prevail on the whole extent of this coast, excepting between Ras-Seger and Ras-Karwan; and in the great bay of Koorya-Moorya, where the weather differs much from that of the rest of the coast. Sudden changes in the wind are often found there, with a violence that is occasionally dangerous, especially as there are no indications of it.

A cloud appears over the mountains of the bay, the fine clear weather becomes suddenly obscured, and five minutes afterwards it blows heavily from the northward. The Arabs call these winds *belaats*, and dread them very much.

Belaats.—During the *belaats* one breeze follows another at intervals of eight or ten days, blowing heavily, sometimes in gusts and squalls, from N.N.E. to N.N.W. These winds never last less than three days, and sometimes continue for seven and even ten days. They are equally dangerous to vessels near the coast. Sometimes towards evening the wind falls entirely, and a calm lasts for an hour or two; then down come heavy gusts from the mountains, at intervals of several minutes between, without any other warning than the whistling of the wind and the noise of the sea made by them. These breezes are heavy enough to blow a sail away from the yard, and even to carry away masts and all. They last for five or six hours, and follow each other at greater or less intervals. The belaat is often followed by strong winds from S.E., producing a heavy sea. At other times, in the same bay, strong winds from S.S.W. often last for five days together, but are not dangerous, as they blow along shore.

Northerly winds are not found any great distance out at sea, and appear to prevail only on that part of the South coast of Arabia abovementioned. After passing Koorya-Moorya Bay they follow the direction of the coast and the high mountain chain of the Subban. Southerly winds only appear to blow on this portion of the Arabian coast, rarely extending over the southern part; and the S.W. monsoon only reaches Socotra from the 1st to the 10th of May. Heavy rains then occur; but the S.S.W. winds of February and March, must not be confounded with the S.W. monsoon.

From the middle of March to April the winds are weak and variable over the whole coast under consideration, and have a tendency to draw to the southward. Alternate land and sea breezes are then found. The sky is generally clear at this time, and the nights are attended by a heavy dew.

In May the weather is uncertain, and when the monsoon is strong it comes from S.W., with fine weather.

In June, July, and August, the S.W. monsoon is in full force; it is especially violent in July, at least it is then most frequent. In the beginning of June the passage is made from the Red Sea or Persian Gulf; but the end of August is preferred, as then the height of the monsoon is over.

In September the wind is moderate, varying from West to South. The Arabs call these *damauro*. These winds continue from the 1st of September to the establishment of the N.E. monsoon.

In October light and variable winds prevail, with frequent calms. The land breeze is found near the coast, and sometimes alternately with the sea breeze. During the night the sky is cloudy, sometimes with rain.

From the 18th to the 20th of November the N.E. monsoon is generally established on the South coast of Arabia. After this period, the winds, taking the direction of the coast, prevail between North and East.

Previous to the monsoon the weather is the same as during the preceding month, and is also rainy.

On the South coast of Arabia the regularity of the seasons must not be depended on, for sometimes at the same period in one year the winds will be quite opposite to those of another. Besides, on the whole of this coast, especially where the land is low. the wind is subject to the effects of the sun, and hence atmospheric changes, an influence even which operates on strong winds. Thus to the N.E. of Ras-Fartak the N.E. monsoon frequently lasts only three months, during a part of November, December, January, and part of February. The S.W. monsoon generally terminates in the middle or towards the end of August, and merchant ships sail for the Red Sea in the beginning of September. In March southerly winds change sometimes to E.N.E. off Ras-Fartak, and are strong in Koorya-Moorya Bay. April and May are considered by the Arabs as a separate season, which they call *bayn el autern*. The wind is then changeable and principally southerly, so that vessels come from all parts. Almost all those which go down the coast at the commencement of the N.E. monsoon, return during these months. In August, when it blows fresh from S.W. to southward of Ras-el-Gat, to the northward of it this wind gradually becomes S.E. From this cape to Muskat variable winds and alternate land and sea breezes are found. Sometimes a N.W. wind, coming from the Persian Gulf, extends as far as Muskat, and occasionally S.E. winds lasting a few days.

In September and after to the end of March, the prevailing wind is East and moderate. The wind is seldom strong at this season, but when alternate land and sea breezes prevail, they are tolerably fresh.

The S.W. monsoon begins at the end of March and lasts till the middle of September.

Persian Gulf.—N.W. winds prevail during all the year in the Persian Gulf, and a southerly wind is only met with in December, January, and February, but even then does not last.

With the exception of these three months, when heavy squalls from S.S.W. and S.W. occur, southerly winds are only occasionally met with in the Persian Gulf. When they freshen up and last two or three days, it is nearly certain that the N.W. winds will then return with more violence than before. It will even happen that in those months, when southerly winds generally prevail, vessels will suddenly encounter heavy squalls from the northward, and perhaps become disabled. It has also been observed that the N.W. winds which follow these southern breezes, are generally stronger and last longer than the former.

Shemaal.—The North winds are called *shemaal.* Those which occur one in a year, in about June or July, and last forty days, interrupted only by intervals of calm and light airs, are called the great *shemaal,* at which time it is useless to attempt beating up the gulf from the southward.

In March and April there is also a little *shemaal,* which sometimes lasts for twenty days without any change in its force or direction. About four or five days before it sets in, the currents of the Persian Gulf set rapidly to the northward, so that vessels beating to windward make about twenty miles a day. The winds in the gulf are neither so regular nor continue so long as those in the Red Sea. They frequently interrupt each other, the land breeze or a strong S.W. wind, principally at the entrance of the gulf.

The N.W. and S.E. winds are those which last; all others are always weak and variable.

During the winter months, southerly winds are accompanied by rain.

In the southern part of the Persian Gulf in October, the wind is variable; sometimes from West and South, accompanied with fine weather. In November the wind varies from N.E. to S.E. and S.W. At the time of the new moon there is generally a strong N.E. wind continuing for three days, with a cloudy sky, after which it becomes very fine. In December moderate northerly winds are found, freshening only at the time of the new moon, at which time it blows fresh from N.E., and the weather is fine and pleasant. In January the wind varies between N.E. and West; prevailing mostly from North and East, and blowing fresh, at the time of the new moon from N.E. Snow then appears on the mountains.

In February the wind generally varies from N.N.E. to S.W., perhaps to N.W. Occasionally there is stormy weather and strong land breezes from the coast of Arabia. There is little rain on this coast, but on the coast of Persia it falls in abundance.

During the month of March, strong breezes prevail from North and West, accompanied with rain, particularly about the time of new moon.

In April, moderate westerly winds prevail, but attended with squalls, which at the time of new moon become stormy. Towards the end of this month we sometimes find easterly winds, lasting for two or three days.

During the month of May the wind is from West and South, bringing fine weather. At the end of this month fogs are very thick, and the atmosphere is loaded with sand, which is kept at a great height on account of its rarity, a phenomenon which especially occurs in the northern part of the gulf and along the coasts bordering the desert.

In the months of June and July the great shemaal is strong from N.W. In August and September the wind is variable, with a tendency to West.

Arabian Gulf.—In the Arabian Gulf, between the coast of Arabia on the West, those of Scinde, Guzerat, Concon, &c., on the East, and the coast of Persia on the North, the S.W. and N.E. monsoons are found; the former with more strength and constancy than the latter. The S.W. monsoon sets in about the beginning of March on the coast of Africa North of the equator. Towards the end of this month it prevails on the coast of Arabia, and in the entrance of the Red Sea, and as far as Ras-el-Gat. In the beginning of April it is found on the coast of Persia, and beyond that gulf; towards the middle of April at Cape Guadel, on the coasts of Scinde, Guzerat, and as far as Cape St. Jean. At Cape Comorin the S.W. monsoon prevails about the middle of April or the end of May, and in the gulf generally about the 15th of June at the latest, and it begins and ends near the coast sooner than out at sea.

From November to December off the coast of Bombay, fresh N.E. winds are found, increasing in strength to the westward. During these two months, as in those of January and February, they veer a good deal to the northward, and bring fine weather.

In January and February the wind is more moderate than in the preceding months; if it becomes N.E. it is interrupted by sudden squalls from South to West, against which the seaman must be guarded.

During the months of March and April, these N.E. winds are not so regular as in the preceding months; they are sometimes interrupted by N.W. winds and calms. But generally speaking in February and March strong breezes, varying from North to W.N.W., particularly in April, are found, and announce a change of monsoon.

Off the coast of Bombay and near this port in March, April, and May, the wind varies from N.W. to S.W., and in the offing to about 6° or 7° West of the coast of Hindostan they incline much to West.

Coasts of Concon, Canara, and Malabar, or the Coast of Hindostan.—On the coasts of Concon, Canara, and Malabar, the West coast of India or Hindostan, the S.W. monsoon prevails at Cape Comorin between the end of April and the middle of May. This monsoon does not reach Bombay till a month later, The N.E. monsoon follows it, and is found on the coast sooner than at sea.

On the coast of Hindostan the N.E. monsoon is neither constant in force nor direction. It prevails at Bombay and on the coast to the northward about the beginning or middle of November. It soon reaches Cape Comorin, Cochin, Calicut, and Mangalore, attended with fine weather. The change of monsoon on this coast is generally accompanied by bad weather: the wind sets in from the S.E., blowing for some hours with a great deal of rain, and after shifting to all points of the compass, inclines towards S.W. This kind of weather, attended by squalls, is generally found about the end of October or beginning of November, and if it does not take place with the return of the N.E. monsoon, it may be looked for in December. In October and during a great part of November, the navigation of this coast is difficult either way, for the wind does not become settled till the end of November or beginning of December.

From November to March, when the N.E. monsoon is well established, splendid weather is found on the West coast of Hindostan, with moderate or fresh winds varying from N.E. to N.W. During these four months alternate land and sea breezes are found along the coast. The first are strong in December and January, and South of Calicut they extend to a tolerable distance in the offing, and sometimes last for twenty-four hours without interruption; but in November and the first part of December they become light. From the middle to the end of February, the land breeze loses force, and becomes irregular. When these alternate breezes are well established, the land breeze begins at six or eight o'clock in the evening, sometimes at ten o'clock, blowing from N.E. to E.S.E. The sea breeze begins about eleven or twelve o'clock in the morning, varying from W.S.W.

to West and N.W., moderate, and at the time of change there is always a calm.

These winds prevail along the coast from Cape Comorin as far as Surate, and cease in the month of March. In April they are succeeded by winds varying from N.N.W. to W.N.W., which continue till the return of the S.W. monsoon.

In March and April the land breeze is very light and of short duration from N.N.E. to N.E.b.N. In fact, during these two months on the West coast of India there is in reality no land breeze, but only winds varying from N.N.W. to W.N.W., which beginning towards noon continue through the first part of the night, falling towards the morning. Sometimes, however, a light land wind springs up, although more generally from midnight till noon light breezes from N.E. and North prevail. At the time of the syzygies the N.W. winds are very strong, and cause a heavy sea.

The weather is uncertain and threatening during May, frequently with rain. The wind is South and S.W., and sometimes strong, especially near Cape Comorin, where the monsoons are found sooner than on any other part of the coast. When the wind veers to N.W., which is the prevailing wind during this month, it brings fine weather. In the nights of May and June the coast is visited by heavy squalls from the high lands. At the end of May, and about the time of the syzygies, or the first part of June, the S.W. monsoon almost generally returns. Its advent is announced by heavy clouds, with a squall from S.E. varying to S.W. and stormy weather for several days. After this the monsoon sets in from South to

S.W. On the northern parts of the coast of Hindostan during this monsoon the wind varies from W.S.W. to S.W. The monsoon commences sooner at Cape Comorin than at Bombay, where it is not established till fifteen or twenty days later. During this monsoon squalls and rain with a very heavy sea are found on this coast, and May is one of the worst months for navigation. In June and July there is some improvement, and by the end of July it is still better.

In August the weather is rather finer, the monsoon not so strong as in the preceding month, and there are bright sunny intervals. The bad weather comes from W.N.W., chiefly on the South part of the coast, as well as N.W., which is the prevailing wind at this period near Cape Comorin.

In September the weather brightens and some fine days occur with the winds from West and W.N.W. This month, however, is not free from squally weather, the wind is not so violent, but the weather is mostly rainy and stormy, with winds varying from S.W. to W.N.W. There is a heavy swell from W.S.W., principally in the intervals of calm between the squalls.

In October the weather is much the same as in September, till the time of the syzygy, when it becomes uncertain on account of the change of monsoon.

At the end of October and November, when the S.W. monsoon is light, near the coast, as above observed, alternate land and sea breezes are found.

The Laccadives.—The Laccadives are subject to the influence of the monsoons prevailing on the adjacent coast. It has been observed that between the South coast of Ceylon and these islands, during the S.W. monsoon, the winds are almost constant from this direction, that they only vary from W.S.W. to S.S.W., and rarely veer to South. These winds are tolerably fresh: they increase in force, and stormy weather is found as these islands are approached. The end of September and beginning of November must be excepted, periods when the winds are light but interrupted by storms and rain near those islands.

From the beginning of November to the end of February, a time when along the coast of Hindostan land and sea breezes alternately prevail about these islands and even to a distance of sixty or eighty leagues from their western shore, the wind is continually from North to N.E., varying three or four points in the twenty-four hours, and even veering to East in the night.

In March and April at the same distance from the shore the wind is from North and N.N.W., varying from two to four points in the twenty-four hours, but generally coming from N.N.W.

In May these winds are variable, and sometimes from West, when the S.W. monsoon is about setting in. When this monsoon is established, S.W. and West winds are found off the coast, with a heavy sea. This weather, however, is not so bad as that near the coast. The storms are less violent and the rains not so heavy.

In September, when the S.W. monsoon slackens, calms prevail near the islands, and off the coast N.W. winds are found, which last till the N.E. monsoon is established. This latter sometimes does not begin till November.

The Maldives.—In February, March, and April, and especially in these two last months, among the Maldive Islands we find N.W. winds varying to W.N.W. The dry season is from October to April, though rain sometimes falls in October. During this season East winds prevail, and are very steady; but according to some seamen in the N.E. monsoon they are only light.

In March the wind from N.N.E. and a S.W. current are often found well to the westward of the Maldives.

CHAPTER IV.

MONSOONS OF THE GULF AND BAY OF BENGAL, AND ON THE SHORES OF INDIA FROM CAPE COMORIN TO THE STRAIT OF MALACCA, INCLUDING THE ANDAMAN AND NICOBAR ISLANDS.

Island of Ceylon.—The island of Ceylon, at the southern extremity of Hindostan, with its West coast following the same direction as that of Malabar, and its eastern coast that of Coromandel, may be considered each as an extension of the continent, and sharing the same winds and weather as are found on its coasts. The two monsoons succeed each other and last about six months. At the end of October the N.E. monsoon takes the place of that from S.W., the seasons change at the same time. On the West coast of Ceylon the N.E. monsoon brings the fine season; while on the East coast it is accompanied with storms and a heavy sea, and either continual rain or dark cloudy and foggy weather. Such weather prevails at this time of the year between the Basses and Point Palmyras, while fine weather is found between Galle and Jaffnapatam, a town situated in the bay of Palk.

At the end of December and in January variable

winds are found throughout a space of fifteen leagues round Ceylon.

On the West coast of this island, and as far as the point de Galle, strong West and S.W. winds are found about the end of March or beginning of April, and last till the month of October. They are generally stronger on the North than on the South part of this coast. The N.E. monsoon follows with fine weather.

On the South coast of the island West winds prevail more or less throughout the year, for even when the N.E. monsoon is in full force, we find alternate breezes regularly established here. Those from the land during the morning come from East to N.E., and towards noon those from the sea from West or S.W. Then North of Cape Basses we find a N.E. wind, which is generally pretty strong, on the East coast of Ceylon.

On the South coast in October and November, strong westerly, varying to N.N.W., winds are found. Heavy stormy weather, interrupted by calms and light winds, or fresh sea breezes, also is found here. In December N.N.E. winds prevail at point de Galle, varying to E.N.E. Off the coast they incline to East, and at this time the monsoon is in full force.

The S.W. monsoon begins on the East coast of Ceylon in February or March; it does not, however, reach to the northward of this island before the beginning of April or May. About the end of March at Trincomalee and the promontory of Palmyras, the weather is then squally with rain. Storms from N.W. occur in the end of April, and at intervals during the monsoon. In April over all the East coast of Ceylon the wind is generally variable and moderate. It is from East to S.E. during the day, and inclines South towards evening; during the first part of the night, the land breeze is from S.W. and West, alternating with the sea breezes. The S.W. monsoon prevails on the N.E. coast of this island about the middle or towards the end of this month, about fifteen days sooner than on the coast of Coromandel. In May we find nearly the same winds as those of April, being from S.E. to South during the day, and from S.W. at night. After the middle of May the sea breeze is at an end, but fresh and sometimes very strong winds blow constantly between S.W.b.W. and S.W.b.S., bringing fine weather.

In June, July, and August, and even to the middle of September, westerly winds blow without interruption. In the month of August, however, on the N.E. coast, a sea breeze, varying from East to S.E., sets in towards noon. It ceases at night, and the West winds then take its place.

Palk Bay.—In Palk Bay, during May, June, and July, S.S.W. winds are violent, and at the change of monsoon on the East coast of this island, as before stated, alternate land and sea breezes are found.

On the East coast of Ceylon in September and October the same winds are found as on the coast of Coromandel. In October, November, and December, rain falls only in the North part of the island. The other months are attended by fogs and very heavy dews.

During the N.E. monsoon on this coast the weather is fine, with fresh breezes from North and N.b.E. In December and January it blows fresh from the northward.

Gulf of Manar.—The N.E. monsoon only extends as far as point de Galle. At the same time, in the gulf of Manar the wind is constantly from N.E., and sometimes very strong. In this gulf during the monsoon it is said that at about thirty miles South of Cape Comorin the atmosphere is free from cloud or mist, a condition which indicates the limit of rain and bad weather. But this is doubted, and it is considered that on the West coast of Ceylon the same weather is experienced as on the Malabar coast.

Gulf of Bengal.—The gulf of Bengal is the sea bounded on one side by the coast extending from Ceylon to Calcutta, and on the other by that between the Ganges and the strait of Malacca. The bay of Bengal is considered the northern part of this gulf. On this extent of sea the winds vary much; and their importance renders it necessary to consider them minutely. Before referring to those of its coasts, the following general observations may be made.

The two monsoons from S.W. and N.E. succeed each other regularly in the gulf of Bengal every six months.

S.W. Monsoon.—The S.W. monsoon commences in the northern sooner than in the southern part of the gulf; it prevails near the coast of Orixa in the beginning of March, and fifteen or twenty days later on that of Golconda; it does not reach the coast of Coromandel till the end of March. On the N.E. coast of Ceylon it prevails, as already stated, between the middle and end of April. But in the end of May the S.W. monsoon prevails throughout the gulf, and is even found as far as the equator.

N.E. Monsoon.—After six months the N.E. monsoon takes its place, and prevails within the same limits by a similar progression; for the S.W. monsoon ceases *sooner where it is later* in becoming established.

Change of Monsoon.—At the change of the monsoon in March and October, the wind is variable; however, in the gulf of Bengal, as well as in the bay, it is more frequently from N.E. than from any other quarter. Violent storms take place at this period, and are stronger on the western than on the eastern coast of the gulf. They are also more frequent and more formidable in October and November than in April, May, or June.

Storms in the Gulf of Bengal.—In that part of the gulf between the coast of Coromandel and the Andaman Islands, storms in October and November come from W.S.W., varying to W.N.W. and even to N.W.; they sometimes come from S.E., but more from N.E. They occur in January from N.E., but are not so heavy.

The storms met with from April to June always begin from N.N.W. or N.N.E., the wind shifting to N.E. and East, and falling when it veers southward of East, blowing hardest from about E.N.E. These storms are accompanied by much rain. Sometimes the wind changes suddenly to S.E. or South, and ceases after blowing hard from S.W., from which quarter it also blows hardest. These however seldom occur. Near the coasts the storms are announced by a heavy swell about twelve hours *before* they take place.

In the middle of the gulf of Bengal, and near its eastern coasts, towards the strait of Malacca, calms are frequent, especially in February, March, and April.

Bay of Bengal.—In the bay of Bengal, and especially in the northern part of it, the regularity of the monsoons is lost, and the winds are very unsteady. The change takes place in April and October. Sometimes the S.W. monsoon changes in the end of September, and sometimes not till the beginning of November.

Storms in the Bay of Bengal.—The squalls of October and November are heavier than those of May or June, the months following the change of the other monsoon. They are also stronger on the western side of the bay than on the eastern, and occur frequently in June. In April, May, and June they are preceded by the S.W. wind falling, and the wind coming successively from all points of the compass, interrupted by calms; the sky near the horizon is clearer than usual: a hollow sound is heard in the rigging, a light wailing sound, well known to seamen, and the gossamer thread is seen about it as in the West Indies, the sea all the time perfectly smooth. The wind is strongest from W.S.W., changing to W.N.W. and N.W., where the wind becomes steady and blows very hard. These storms last ten or twelve hours.

Off Cape Palmyras these storms generally commence as above described; sometimes however the wind changes from N.W. to North to N.N.E. and to N.E.; a calm then ensues, lasting a quarter or half an hour, followed by a S.W. wind as violent as the former. This wind goes down after lasting about an hour, and is followed by the wind from E.S.E., varying to East and N.E.; it then falls and commences again from S.W. as before. These latter storms are much less violent than those in which the wind is from S.W., but they are always attended by a deluge of rain.

The storms of October and November are heavier than those in the months we have mentioned. They are preceeded by a slight rain and the wind from E.S.E., which gradually increases in force, changing to East, E.N.E., N.E., and North, at which pcint it subsides. Sometimes in very violent storms the wind changes to N.N.W. and N.W., attended by heavy rain. In these months storms also occur from East and E.N.E., varying to North, and lasting one or two days. They are followed by a calm of some minutes, and the wind then returns with violence to S.S.W., subsiding in about half an hour.

Lateness of the N.E. Monsoon in the Gulf of Bengal.—The N.E. monsoon should begin in October in the gulf of Bengal, but it is rarely found in this month in the southern part of the gulf. Between Ceylon and the strait of Malacca, from the equator to the parallels of 8° and 10° N. latitude, in October and November strong westerly winds are found to last several days. Near the equator the wind is generally from N.W. and N.N.W. On a line from Ceylon to Cape Achen they vary from W.S.W. to W.N.W., while in the North part of the gulf they are from S.W. to S.S.W. In October and November these winds most frequently prevail near the Nicobar Islands, and in the space separating these islands from Ceylon. In making for Ceylon or the coast of Coromandel during these two months, a ship must not approach Cape Achen.

During the N.E. monsoon in the Gulf of Bengal the

wind varies from N.N.E. to E.N.E. During the S.W. monsoon it comes principally from South. This monsoon ceases sooner out at sea than on the coasts.

S.W. Monsoon in the Gulf of Bengal.—In March the monsoon varies from S.W. to South, inclining eastward, and even to N.E. Southerly winds prevail near the coast; while out at sea the wind is N.E., varying sometimes to S.E., but rarely to S.W.

In April and May the wind in the middle of the gulf is from S.S.W. to S.W., but in May is changing to W.S.W. Sometimes in April it is moderate and variable, with occasional calms, dropping from N.E., varying to East and S.E.

In June, July, and August, the monsoon is at its height, and steady between S.W. and West, especially in the two former months. About the end of the latter, and in August, it often veers to W.N.W. and even to N.W.

N.E. Monsoon.—In September, October, and sometimes even from the middle of August, the wind is from West, veering to S.S.W., accompanied by heavy rains, which last until the beginning of October. After the 15th of September and in October the S.W. monsoon becomes weaker, the wind variable and frequently from N.E. and N.N.W.

In November the wind varies from N.N.E. to E.N.E., as well as in December and January, months when the monsoon is well established in the gulf. October and November, however, are stormy months, as already stated, and bring an abundance of rain. From the middle of December to the end of January the N.E. wind is not so strong as in the preceding months.

February is the finest month of the year, with a fresh breeze from N.N.E. to N.E. At the end of this month the N.E. monsoon is failing, and southerly winds are more frequent than those from the northward. About the end of February in the southern part of the gulf they are from South and S.S.W.

In January, February, and March, the N.E. winds often veer to S.E., but seldom to S.W.

Coast of Coromandel.—On the Coromandel coast the S.W. monsoon, which prevails from the middle of April to the middle of May, is the fine season. The N.E. monsoon, which succeeds it, is accompanied with rain and stormy weather. The storms, however, are not so frequent or formidable as on the Malabar coast.

In March the weather is fine, storms seldom occur and are of short duration. In the fine weather of this month, which is about nine-tenths of it, the wind is moderate, varying from S.W. to S.E. In the afternoons the S.E. wind prevails, inclining sometimes to East, and even, although seldom, to N.E. After midnight the wind will be from N.E., varying by N.W. to S.W., sometimes followed by a calm; but when fresh it lasts till nine or ten o'clock in the morning.

In April the N.E. monsoon ends. The alternate land and sea breezes, however, still prevail at times during this month. Those from the sea rise towards noon and last during a greater part of the night; they then change to S.S.W. and S.W., blowing fresh. The S.S.E. winds are also strong, and often begin about nine or ten o'clock in the morning, sometimes lasting all night. Towards the middle of the month the wind becomes variable, the weather uncertain and sometimes rainy. The southerly winds, changing from S.W. to S.E., still prevail, these last often blowing strong. When, towards evening, they incline to the southward and S.W. they become weak, occasionally falling calm. This, however, is not often, and they are generally followed by a squall or the return of the N.E. monsoon. In this month no leewardly ship can make southing on the Coromandel coast on account of the strong N.E. current. In the month of April about twenty-four fine days are reckoned on and storms are few, although the weather frequently assumes a stormy aspect and lightning is seen in the West. Still, fresh breezes, accompanied with rain, do occur and last two or three days,—a time when a ship should give the coast a good berth, not standing into less than fifteen fathoms on account of the heavy sea on. At the end of April westerly winds prevail and continue some days, while at sea the wind is light, variable, and interrupted by calms.

In May the wind varies between E.S.E. and S.W. on the Coromandel coast. The sea breeze rises at noon and blows fresh from S.E.; it sometimes lasts till midnight and changes to South and West, blowing till morning. Sometimes one wind follows the other, blowing equally fresh, but at other times there is an interval of calm between them. From the middle of April to the middle of May storms sometimes occur.

Storms on the Coromandel Coast.—The stormy weather on the Coromandel Coast generally begins from N.W., the wind changing to North, N.E., East, or E.b.S., at which point it goes down. In these storms the wind is very heavy from N.E., but when it veers to S.W. by the South it is much more so. These tempests are generally preceded by a heavy swell rolling towards land. Some storms are merely attended with rain and little wind. Such storms as the foregoing are not of common occurrence, and when such a one is at hand a ship should get to sea while the wind is from N.N.W., before it veers to the eastward, from whence it blows hardest. About two-thirds of this month is fine, with a clear sky. In the evening, however, it often clouds over from the West, and when the wind is fresh during the day the weather becomes foggy and the horizon misty.

In June, July, and August the S.W. monsoon is in full force, the wind varying only from W.S.W. to West. If it be moderate, the sea breeze, which is light and very uncertain, blows from S.E. to South, taking the latter direction during the evening. It begins at 1h. p.m. and falls about nine or ten in the evening, sometimes not before midnight. In the morning the wind shifts to S.W., sometimes to West, and blows strong. The West and S.W. winds in these three months are sometimes hot and parching—indeed almost insupportable. They often blow with such violence as to darken the sky with the dust and sand which they raise. These clouds of dust are sometimes carried far out to sea, especially during the dry season. In August the frequent rains prevent the wind from producing this effect. In this season, while the weather is fine on the Coromandel coast, heavy rain is falling in the bay of Bengal, as well as on all the East coast of the gulf. At Balassore, even, the W.S.W. winds are very strong and disagreeable.

In the month of August the steadiness of the S.W. wind is not so evident as in the preceding months. The land and sea breezes are more regular; the latter coming from S.E. and towards evening shifting to South and S.W. In this month squalls occur from S.W., with rain, and are sometimes heavy. Towards the end of this month the wind is from W.N.W., hankering to N.W. It happens sometimes in August that a strong N.E. wind springs up for one or two hours, bringing rain in abundance. It afterwards veers to E.S.E. and South, blowing strong all the while, and then shifts back to East in squalls, still accompanied by rain. But when these winds shift to S.W. the weather again becomes fine.

In June there are about eight days fine weather, eleven days cloudy, and eleven overcast. July has about eight fine days; the others are generally overcast or cloudy. In August lightning occurs without thunder, and, as in June, there are not more than eight fine days.

In September both wind and weather are variable; sometimes stormy and rainy, West being the prevailing wind. This varies between W.S.W. and West, but sometimes as far as N.E. In the day the wind is from N.E. and, perhaps more frequently, from S.E. to S.S.E. During this month the wind is generally very moderate, from whatever quarter it comes, and if a storm or two does occur in the latter part of it, when they come off the land they are very brief. The sea breeze gets up about two or three o'clock in the afternoon, and lasts till eight or ten at night. There is much lightning, especially in the evening, but seldom thunder. The weather of August is very similar to that of September.

N.E. Monsoon.—The N.E. monsoon generally commences between the middle of October and the 1st November. In October the clouds become heavier than in the preceding months. Calms, lightning, and rain are frequent till towards the 9th of the month, the winds being variable, with frequent squalls. Sometimes in the morning the wind is East. At other times it is very fresh from N.N.W. for several days; it then comes in squalls with rain, in the afternoon shifting to N.E. or E.N.E., in which case it brings rain. However threatening the appearance of the weather, storms of any violence seldom occur. It is not right, however, to remain on the Coromandel coast beyond the 15th or 20th of October. Towards the end of this month the weather is generally bad and squally, attended with heavy rains, thunder, and lightning.

In November the wind in the morning is generally N.E. to N.N.E., varying to N.N.W. and N.W. After noon it becomes N.N.E. or N.E., sometimes shifting to S.E., and even to S.W., for several days, a circumstance, however, which is very rare. Calms last several days in this month, followed by storm and rain. The N.E. monsoon may be introduced by a heavy breeze from N.W., which changes to North and N.E. and ends at East; but this monsoon sometimes begins from N.E. In the middle of the gulf of Bengal it blows from S.W. and West. From the beginning of October to that of December stormy weather is common. In these storms, when the wind shifts to the southward of East it falls and the weather clears up. If, on the contrary, after blowing hard from N.E. it falls calm, it afterwards shifts to S.E., to South, and S.W., and again blows hard. Nearly the whole month of November is dark and gloomy.

In December the N.E. monsoon is well established, and the rain not so heavy. In the morning the wind gets up from N.N.W. and N.W., veering in the afternoon to N.N.E. and even to E.N.E. When it continues at N.E. it becomes strong and blows for several following days. The weather is finer and more settled than in the preceding months, and about thirteen really fine days are looked for in December.

In January and February the weather is fine, the sky serene, and the sea smooth. N.E. squalls are rare, and, if any, of short duration. In the day the wind is generally from E.N.E. to N.E., and at night it becomes N.N.W. or N.W. In February the N.E. monsoon weakens; it lasts, however, till the middle of the month. Calms then succeed, and frequently S.E. winds. The monsoon lasts longer out at sea than on the coast. In some years North winds last till the beginning, and even throughout the month of March. While N.E. winds are prevailing in the gulf of Bengal, on the Coromandel coast the wind is South; and while the wind is South in the gulf it is N.E. on the Coromandel coast. In January the weather is very calm, and twenty days of fine weather may be reckoned on. But February is considered the finest month in the year, and in this month rain and lightning are most rare.

Coasts of Golconda and Orixa.—On the coast of Golconda, from the middle to the 20th of March, the S.W. monsoon, which brings fine weather, is found on the coast near Masulipatam. It prevails on the coast of Orixa about the beginning of that month, and therefore this monsoon begins in the head of the gulf sooner than in the southern part. The N.E. monsoon follows it at an interval of six months, commencing at corresponding periods. This is the bad season, especially in November and December.

On the coasts of Golconda and Orixa in the end of February the wind is from N.E., sometimes fresh and sometimes light. The land and sea breezes generally begin in the early part of March, those from W.N.W. varying to S.W. In March the wind is from S.E., hauling to E.S.E. during the day, and from S.S.W. to S.W. during the night. Sometimes strong winds from N.W. are found on these coasts, and also at Balasore, West of the mouth of the Ganges. These are most common in the end of March.

In April and May the wind is frequently strong from South, especially on the coast of Orixa. It varies from South to S.E., bringing cloudy and misty weather. In May it changes from S.S.E. to S.W., and sometimes to W.S.W., on the coast of Orixa and Balasore. In May it is sometimes strong from S.S.W. to S.W., with cloudy weather: a season of the year which is very bad and stormy, especially in June. From the end of May till the end of June it is dangerous to go to Balasore, and also between the end of September and the end of October, at the change of the monsoon.

In June and July rainy squalls are frequent. In June, July, and August the S.W. monsoon is in full force on the coast of Orixa, the sky overcast.

At Balasore the S.W. monsoon blows fresh from the West. In April and May it is often South and S.E. If it shifts from S.S.E. to S.W., and blows hard, it brings rain, while the weather is fine on the southern part of the coast. In June, as well as July, it blows from South, veering to S.E. In August the weather is uncertain; the wind also varying from S.S.W. to W.N.W., and sometimes coming from S.S.E., attended with heavy rains and squalls near the coast of Balasore.

About the 21st of September on the coasts in question we have strong easterly winds, and the end of the S.W. monsoon is distinguished by tempestuous weather.

In October and November the N.E. monsoon is established. In the former month, however, the N.E. winds are variable. November and December are two months of bad weather, coming in sudden squalls.

Towards the middle of January the N.E. wind falls, and towards the end of it the rain ceases. Land and sea breezes then set in, the former W.N.W., the latter E.N.E.

Bay of Bengal.—The bay of Bengal, as already said, may be considered the portion of sea to the northward of the gulf of that name. It is bounded on the West by the coast of Orixa, which we have just been considering; on the North by Bengal, and on the East by the coasts of Chittagong, Aracan, Ava, and Pegu.

The S.W. monsoon, with the wind from South and

abundance of rain, prevails in the bay during the first part of March; the N.E. monsoon prevails six months afterwards. During the latter it varies from N.N.E. to E.N.E., and according to some from N.E. to N.N.W.

Besides the squalls already mentioned as met with near the coasts from April to August, when the S.W. monsoon is in full force, there are others of less duration, which intervene at different periods of the year, principally off the coast of Bengal. The squalls during the S.W. monsoon sometimes come from S.S.E., but more frequently from South to S.W., and sometimes from West. In September, October, November, and even the beginning of December, squalls from the southward are found, but they seldom occur in the latter month.

In March and October, when the monsoon changes, the wind is most variable, blowing more frequently from N.E. than from any other quarter. This is the stormy season. When the monsoons become weak, land and sea breezes are regular, but those of the end of the S.W. monsoon are not so strong nor so regular as those about the end of the N.E. monsoon.

The change of monsoon at the entrance of the River Hoogly is often indicated by a N.W. wind, coming in gusts; winds which are announced by thick clouds rising suddenly, and sometimes accompanied by lightning. But they are sometimes very violent.

Towards the end of March or beginning of April the S.W. wind becomes fresh and regular, varying to S.S.W. and W.S.W. The weather is then cloudy, dull, and rainy. But during the last half of May storms are rare. Sometimes in March the wind varies from S.W. to S.E. and even to N.E.

April and May, according to some seamen, are the most dangerous months in the bay of Bengal. In this last month we have storms from North; generally, however, we find June and July the worst months for navigation. From April to June the wind generally varies from South to S.E.; in June, July, and August, from S.W. to West, and sometimes in this latter month from W.N.W. to N.W. In August, notwithstanding bad weather may be expected, there are some days of fine weather.

In September the S.W. monsoon begins to fail, the wind is moderate fron N.W.. and rain continues to fall abundantly. The winds in this month are changeable, the weather sometimes fine, but at other times rainy and stormy. After the 15th of September, in crossing the bay of Bengal for the coasts East of Balassore, the wind is found varying from N.E. to N.N.W. Towards the equinox the wind is strong from East in the environs of Balassore, and after this period in the middle of the bay N.W. winds are found. Before the change of the monsoon a short but strong gale of wind prevails over the whole extent of the bay.

In October the rains cease, after falling in abundance on the coasts of Bengal, Chittagong, Aracan, Ava, and Pegu. In the beginning of this month we still have southerly winds, interrupted by variable and strong breezes from N.E. to East. As a general rule it may be considered that during this month the wind varies from N.E. to S.W.

In November the wind is strong from N.E. From this month, and sometimes from October to March, the prevailing winds are from N.N.E. to N.E. In December and January the winds are nearly the same as in November. In February, which is the finest month on the eastern coasts, the wind is nearly always between North and East, while on the West coast land and sea breezes are regularly established, as well as fresh breezes, varying from W.N.W. to S.W. The land breezes are less regular at a distance off shore. Lastly, it may be observed that the N.E. monsoon commences in the early part of October on the coasts forming the head of the bay of Bengal, and continues to a later period in the middle and South part of the bay and between the islands of Nicobar and Ceylon. In these latter parts the wind from S.W. and West often lasts during the whole of October, and sometimes only terminates in November.

North of the parallels of 17° and 18° the wind is often light during the N.E. monsoon. It is sometimes from this quarter, but generally varies from N.N.E. to N.N.W. In the bay of Bengal calms and light breezes are frequent during the N.E. monsoon, particularly on the North coast and the coasts of Aracan and Ava, and we also find there strong southerly winds. It may, however, be observed that during this monsoon the prevailing wind in the day is from N.W. to W.N.W., and during the night from North, rarely varying to N.E.

Coasts of Martaban and Tenasserim, Andaman and Nicobar Islands.—On the coasts of Martaban and Tenasserim the S.W. monsoon begins between the middle and the end of May. The N.E. monsoon succeeds it six months afterwards. During this latter monsoon, among the Andaman and Nicobar Islands northerly winds are found, while at sea the wind is N.E.b.E. and E.N.E. The S.W. monsoon is not regular from twenty-five to thirty leagues East of the Nicobar Islands. The wind there varies from West to North, sometimes even from S.E. to S.W.b.S., but most frequently it is from N.W. and is attended with squalls and heavy rains. It is sometimes light and variable, with calms, a smooth sea, and a cloudy sky. Near the Nicobar Islands at the time of the change of monsoon tornadoes of an hour or so are met with.

The N.E. monsoon begins in November and brings the fine season. About the Nicobar Islands the wind is light during this monsoon, and often interrupted by land and sea breezes. Here, also, the wind is from the northward, while at sea it is from N.E. to E.N.E. Northerly winds especially prevail to the northward of these islands during December and January. Sometimes at the end of April there are sudden westerly winds. The S.W. monsoon begins in May; the winds at that time near the Andaman and Nicobar Isles being strong from South.

In the islands near the coast of Tenasserim the wind is light in July and August, shifting from North to West. The strong winds which prevail at this time in the bay of Bengal do not reach these islands till they have lost much of their force. On the coast East of Cape Negrais, in the gulf of Martaban, the same fact is observed, and calms are found on this coast chiefly from February to April, but seldom of long duration.

On the coast of Martaban and Tenasserim, from November to January the N.E. monsoon constantly prevails. In October and November sudden winds are met, commencing from E.S.E. and shifting to E.N.E. and North, where they cease, and are attended by heavy rain. In the middle of February, and in March and April, the period when the N.E. monsoon weakens, land and sea breezes prevail on these two coasts; they are moderate, blowing from East to N.W. chiefly in February. On the coast of Tenasserim, nevertheless, the wind is oftener from the southward than the northward. In these three months calms are common on these coasts, but of short duration.

The S.W. monsoon, on the coasts of Martaban and Tenasserim, is a strong wind and sometimes even violent. During the months of May, June, and July, the wind varies very little, its direction being from West to W.S.W. The monsoon is in full force from August to the middle of September, when it decreases, becomes variable, and is interrupted by calms and light breezes.

In the beginning of October, the month when the S.W. monsoon ends, the land and sea breezes again prevail on these coasts, a condition common to almost all the coasts of India in the intervals between the monsoons. These breezes are not so strong as those which follow the end of the N.E. monsoon.

CHAPTER V.

MONSOONS OF THE STRAIT OF MALACCA, COASTS OF SUMATRA, JAVA, TIMOR, AND WESTERN AUSTRALIAN SHORES FROM TORRES TO BASS STRAITS, WITH THE MOLUCCA CHANNELS.

Strait of Malacca.—In the strait of Malacca, although it is within the limits of the monsoons, the wind is very variable; but land and sea breezes are found regular on the West coast of Malacca, as well as on the N.E. of Sumatra; and in the strait the monsoons are only regular when they are at their height in the surrounding seas. Even then, however, the wind is moderate in the channel and only lasts during a part of the day. The N.E. monsoon brings the fine weather, and this season lasts from November to May. That from S.W. generally begins towards the end of April or beginning of May and ceases in October. In November, however, westerly winds are still found, and during this monsoon the weather is generally cloudy and stormy, with rain, especially when it is at its height.

In October and November the wind often changes from N.W. to West; sometimes, however, when from N.E., it blows regularly in November. It is tolerably strong till the month of March, especially in December and January, sometimes varying to North and N.W.; and during the N.E. monsoon, in every month, one or two days of westerly winds occur. During the N.E. monsoon the wind generally varies between N.N.E. and E.N.E. Towards the end of February, in March, and sometimes early in April, the N.E. wind will veer northwards, becoming light and variable. We then find alternate breezes, interrupted by calms about noon, while through the night, and at sunrise, the wind is fresh. The coast of Malacca is less subject to calms in this monsoon than that of Sumatra.

The S.W. monsoon is at its height in June and July. From May to September the winds in the strait are chiefly from S.W. to South,—that is, at the period when, out at sea, the monsoon is at its greatest height. In the course of this monsoon calms are found on the N.E. coast of Sumatra, but by no means so frequently nor so long as on the promontory of Malacca. The middle of the day is generally calm, and a fresh breeze is blowing at night until sunrise. It is only in the northern part of the strait of Malacca that the monsoon is ever well established.

During the S.W. monsoon, in the vicinity of Mount Parcelar and Cape Rechado, the S.E. wind will incline to S.S.E., and about the middle of the night the "Sumatras" will come on from S.W. to South in a fresh breeze from the coast of this island. They are strong winds descending from the mountainous parts of the island, attended by rain and thunder, generally setting in heavily and lasting for two or three hours. Though the Sumatra is met with in all parts of the strait it is mostly found near the coast of Pedir and between the mountains of Parcelar and the islands of Carimon. Here it commences in a heavy squall and lasts from six to eight hours, blowing very hard in successive squalls with intervals of moderate breezes between. In Malacca Roads, where these storms occur between seven o'clock and midnight, great precaution is necessary.

Storms from N.W. are not so frequent as the Sumatras. They are most common in the northern part of the strait, but are sometimes met with near the Carimon Isles and in the strait of Singapore. They commence with a sudden violent gust, which soon slackens, and are announced by a black arched cloud rising rapidly, scarcely giving time to reduce sail, and, like the Sumatras, are accompanied by rain and wind. A vessel, bound to the southward, waiting tide at anchor in this part of the strait, and observing indications of a storm from N.W., should lose no time in getting under way before it comes, in order to take advantage of it and run with it while it lasts, more especially as, if she has it at anchor, she would find it more difficult then to do so, and would thereby lose a favourable opportunity of proceeding down the strait.

On the West coast of Sumatra, which island is divided into two almost equal parts by the equator, we find at the same time different monsoons, according as we are North or South of that great circle. Thus, from October to April, while the N.E. monsoon prevails on the part of that coast which lies North of that line, the N.W. monsoon (which in the South hemisphere also prevails from October to April) is blowing on that part of the West coast situated South of it; and again, while the S.W. monsoon (from April to October) is ėstablished on the N.W. coast of the island, the S.E. monsoon (from April to October), which is only a continuation of the Trade winds of the southern hemisphere, prevails on the S.W. coast. This latter monsoon is always free from storms, which are only met with along this coast during the N.W. monsoon. In this latter, which is generally more violent than the S.E. monsoon, heavy winds, with rain, are met with; while from May to September, the weather is always fine on the S.W. coast of Sumatra.

West Coast of Sumatra, North of the Equator.—On the West coast of Sumatra, North of the equator, the S.W. monsoon sets in without difficulty. The wind is strong from April to November. Towards the end of August it veers to South till the middle of September, varying to S.S.W. near the coast, while at sea it is from W.S.W. to S.W. Near Pulo-Way it is from S.S.W. to South about the middle of September. During the S.W. monsoon the weather is uncertain to the North of Sumatra.

These observations on the winds apply especially to Achen, for when the S.W. winds, which blow in the gulf of Bengal, reach the N.W. part of Sumatra, they seem to be checked by the high lands of this island, and to take a direction along the coast from N.W. to S.E. They bring rain and bad weather; and are, besides, only well established when the S.W. monsoon is in all its force, namely, in June, July, and August. At Achen the rainy season is that of hot weather. It begins in April and ends in August.

The N.E. monsoon, which, on the N.W. coast, follows the S.W. monsoon, brings fine weather, and the wind is less violent than in the preceding monsoon. The N.E. wind, which is very fresh on the eastern coast of the island, is interrupted by the high lands, and is only felt out at sea, at a tolerable distance from the N.W. coast of Sumatra, about thirty or forty leagues North of it or even sixty or eighty leagues, as we approach it from the equator. Sometimes a S.E. wind is found instead.

In December and January alternate land and sea breezes prevail along the coast as far as the equator, and even as far as Bencoolen. The N.E. monsoon is sometimes interrupted, but very seldom, by N.W. winds.

West Coast of Sumatra, South of the Equator.— On the S.W. coast of Sumatra, in the course of the S.E. monsoon, the wind varies from S.S.E. to S.S.W.; sometimes in May, June, and July, it is interrupted by a return of N.W. winds, lasting several days, especially at the time of the syzygies. This also takes place in the strait of Sunda and on Java.

About the equator, in this monsoon, the winds are very variable, and continue so nearly all the year, with occasional calms, but very seldom squalls, especially near Pulo-Mintao and Priaman. In this last part squalls frequently come from the mountains at night.

The N.W. monsoon on this coast begins in November and ends in March, reaching each place on the coast gradually later, according as it is more distant from the equator. From October to April it blows fresh, varying from N.W. to W.N.W., and reaching as far as the strait of Sunda. It is sometimes interrupted by South and S.E. winds, and by squalls accompanied with rain.

In April, May, October, and November, the periods of the change of monsoon, the wind is very unsettled. Towards the end of each monsoon alternate land and sea breezes take place; and it may be observed that the land breeze does not frequently blow in the opposite direction to that from sea.

Where the coast lies North and South, the sea breeze is from N.W. and the land breeze from N.E. Where the coast lies N.W. and S.E. the sea breeze is from South and the land breeze from East. When the wind veers from East to North, the sea breeze may be expected on the following day to occur from West to N.W.

The land breeze does not reach beyond three or four leagues out at sea; it first begins near the coast, while a calm prevails at four, six, or eight miles to sea; it then gradually extends further seaward as the heat of a place is more considerable, and attains its greatest distance from the coast in the course of an hour. At the change of monsoon the sea breeze is generally followed by a land breeze; but these breezes at the end of the S.E. monsoon are not so strong nor yet so regular as the others.

Sunda Strait.—In the strait of Sunda the wind varies from S.S.E. to E.S.E. from April to October, a period when it is called the East monsoon; and it varies from W.N.W. to N.W. during the West monsoon, which follows. This latter monsoon prevails in November and brings bad weather. In this strait there are alternate winds, from South before noon and from North in the afternoon, separated by an interval of calm.

Among the Sunda Isles, as far as Timor, the monsoons are the same as abovementioned. The East monsoon begins in May, the wind varying from East to S.S.E., and attaining their height in June and July. This monsoon is finer than the westerly monsoon, which brings bad weather, especially in November and December. The rains set in during this month, accompanied with squalls. The westerly monsoon begins in November, and attains its height in January. Rain sets in in December, lasting till the middle of February, accompanied by storms and severe weather. Then the monsoon gradually weakens till March. In April the wind becomes variable and the weather is pretty fine.

South Coast of Java.—On the South coast of Java the wind is from N.W. while the N.E. monsoon prevails to the North of the line—from October to April; which monsoon ceases in March. In April the wind is unsettled; in May it is steady from East, with fine weather: and from June to August it is strongest. In October the S.E. monsoon becomes weaker, and until the return of the N.W. monsoon the wind is variable. In the months of May and November the rains on this coast are very heavy.

In February and during the first half of March, as well as in October,—that is, at the change of monsoon,—alternate land and sea breezes prevail; but they are not so strong in October as in February and March. In these two last months, and even in April, the land breeze begins with a squall or sometimes with a heavy storm.

and as soon as it is over the land breeze is found blowing moderately, and continuing until the return of the sea breeze. In April and May, on this coast, the sea breeze also begins with a heavy squall or a storm of short duration. We mention these facts; of which we know no other instances except in the island of San Domingo, where this phenomenon takes place in the roadstead of Port-au-Prince and in the channel of Gonaive.

Island of Timor.—On the N.W. coast of Timor, from September to March, the N.W. monsoon is found, varying to N.N.W. In April or May it is succeeded by that from S.E., varying to S.S.E.; which terminates in October. The N.W. monsoon is the bad weather season, and in December the winds are very violent. This monsoon is only well established here about the end of November or December, and heavy winds, between West and North, accompanied by rain, continue till February. About the end of April or beginning of May the wind returns to East, varying to South, blowing strong on the North coast of this island, where it is then the fine season. The strongest winds vary between W.b.S. and N.N.E.

On the opposite coasts of Timor there is a great difference in the winds. The S.E. monsoon is very light on the South coast while it is very strong on the North. The South coast, during the first part of October, is stormy; while on the North these are only experienced in December. On both coasts during the fine season the land and sea breezes are strong. On the South coast the land breeze varies from N.E. to North, the sea breeze from S.S.E. to S.S.W.

In that portion of sea separating the isles of Sunda from Australia calms are very frequent. Two monsoons are found there. One from West, in which the winds vary from S.W. to N.W., begins in October and terminates in April. During this monsoon it blows fresh, with occasional squalls, from December to February. The other monsoon, from East, varying to S.E., is really a continuation of the Trade wind, and brings the fine season. The zone of the monsoons does not pass South of the parallel of Christmas Island. But to the East of this island its southern limit gradually approaches the parallel of 14°, sometimes even further South, and it generally varies between 14° and 15° S. To the eastward it reaches the N.E. part of the coast of Australia.

Torres Strait.—In Torres Strait easterly winds prevail. The westerly monsoon does not blow there steadily; it is frequently modified by the easterly wind, which is then light and variable, and lasts several days, until it strengthens to a fresh breeze.

On that portion of the sea between Papua or New Guinea and Australia, during the month of January and at the beginning of the westerly monsoon the winds are generally from N.E. to North, drawing occasionally to the westward.

Near the N.E. coast of Australia, as far as the parallel of 14° S., winds varying from N.E. to W.N.W. prevail, and further South they veer to East and E.S.E.

Between these two monsoons there are frequently calms of long duration, and the time of the change from the S.E. to the N.W. monsoon is the period when these long calms mostly prevail. When the monsoon is about to be established westerly winds blow for five or six days, then they cease, and are sometimes succeeded by light variable winds for a month. Then at the following syzygy the monsoon becomes established, with gloomy rainy weather and sometimes squalls, for two or three days. The weather then clears up and a moderate breeze sets in for some time, producing clearer and finer weather than is experienced during the S.E. monsoon.

Two or three days of bad weather may be expected at the period of the syzygies, although sometimes five or six weeks of continuous fine weather may have prevailed. Near the land the weather is always more boisterous and rainy than at a certain distance out to sea; yet about the limit of the monsoon, in 15° S. latitude, rainy and squally weather is generally met with. The mean direction of the wind is nearly W.N.W., veering to N.W. and S.W. at the time of the syzygies, and sometimes at those periods even to W.S.W.

North Coasts of Australia.—On the N.W. coast of Australia, from Melville Island to the N.W. cape, there are land and sea breezes. The former, from West, vary to N.W.; the latter from S.S.W. and S.S.E. These winds blow chiefly during the summer of this hemisphere; in the other months they are very variable and not so regular. They extend only a short distance from the land. But off the N.W., as well as the western coast, the general wind prevails. Off the western coast of this continent, North of the tropic, we find, like the foregoing, alternate land and sea breezes, but observations are so scanty here that we can hardly arrive at any general rule.

The following table, drawn up at Perth—the capital of the Swan River settlement—in 1832, from Montgomery Martin's *History of the Colonies*, will give a general idea of the climate of this coast.

Months.	Ther. Far.			Barometer.			Winds.	Weather.
	Max.	Mean	Min.	Max.	Mean	Msn.		
	°	°	°					
Jan.	99	80	61	30·1	29·9	29·8	N.E., S.W.	First part fine; mid cldy; 4 cler.
Feb.	106	88	59	30·1	30·0	29 8	N.E., S.W.	Cldy days 8; 3 rain; rest fine.
March	95	79	63	30·2	29·9	29·6	N.E., S.W.b.S.	Genlly fine; rain 4th and 5th.
April	92	72	63	30·3	30·1	29·9	N.E., N.W., S.W.	Like our April.
May	74	63	48	30·5	30·1	29·5	E.N.E., N.W., S.W.	Rainy days 17; N.W. winds.
June	74	59	45	30·6	30·2	29·8	W.S.W., N.W.	Frost 16; ice 3; hvy strm, hail.
July	70	56	41	30·3	30·0	29·6	N.E., N.W., S.W.	Genlly fine; ice 1st & 9th; strm.
Aug.	79	63	45	30 3	30·0	29·7	N.E., S.W.	Rain 10; damp; storm on 11th.
Sept.	76	63	50	30·3	30·0	29·7	N.W., S.W.	Cldy; sqlls & rain; strm on 18.
Oct.	76	63	50	30·2	29·9	29·5	N.E., N.W., S.W.	Alternately fine and foul; sqlly.
Nov.	86	70	52	30·4	30·1	29·8	S.W.	Gale on 1st; squally 9th & 10th.
Dec.	101	81	81	30·1	29·9	29 7	S.W., N.W.	Cloudy, clear, rain; mid. fine.

On this same coast, from the tropic as far as Cape Leuwin, the prevailing winds are from N.W. and S.W. The climate is warm, rainy, and entirely free from snow. The strongest winds are from N.W., and next to them those from S.W.; the former blow with especial violence near Cape Leuwin. Hot northerly winds on the coast are trying, but they generally last only a short time. During the summer months regular land and sea breezes prevail. The former begin at night and cease in the morning, varying from East to N.E.; the latter rise in the afternoon, blowing from West or S.W., and cease after sunset.

Off Cape Leuwin the prevailing winds are from West,

varying during the summer from West to N.W. in the night, and to S.W. in the afternoon. These variations, however, are not regular, and never happen in the winter.

On the South coast, between the Recherche Archipelago and Bass Strait, the prevailing winds, from the middle of January to the middle of April, are from S.E. to E.N.E. During the night they blow from the land, and in the day time from the sea. They are seldom strong; while, on the contrary, the occasional westerly breezes, which also occur at this period, are generally very strong, and sometimes more than that when they veer to S.W. Off the South coast of Australia, in taking a general view of the subject, it may be said that during the six or eight winter months southerly winds are always found, varying to W.S.W.

During this season squalls from S.W. are very frequent. In these the wind gets up from N.W., accompanied by fog or rain; as it increases in violence it generally veers to West, and as soon as it has a little southing in it the weather begins to brighten. As soon as it veers to S.W. it is in greatest force. The barometer, which at the commencement of this squall falls to 27·0 in., and sometimes lower, then begins to rise, and as the wind passes to South and S.S.E. it becomes weaker. The weather then quite clears up and the barometer rises to 28·0 in. It sometimes happens that the wind changes to West and even to North at the same time that the barometer falls. Although the wind may not then be so violent it must not be considered as over. Sometimes it returns after an interval of one or two days. Sometimes also the wind veers rapidly from N.W. to S.W., and then fog and rain last longer. Such in general is the course of the changes of wind near the South coast of Australia as well as in Bass Strait; while to the East of this strait, and on the East coast of Tasmania, the wind is generally from South to S.E., frequently accompanied with rain and fog.

Off the South coast of Australia the barometer rises with South winds and falls with those from North to South by the West; to the East and West of this continent it generally rises with moderate breezes from the sea and falls with those from land.

North winds do not prevail near the coast. They are found between the parallels of lat. 40° and 44° S. To the West of Tasmania strong breezes often found from N.N.E. sometimes shift suddenly to N.W. and West. These winds are found between the abovementioned parallels in the whole space comprised between the bank of Agulhas and Tasmania, their changes being sometimes very capricious.

Bass Strait.—In Bass Strait squally weather and strong breezes come from S.W. and prevail during nine months of the year. In January, February, and March, sometimes easterly winds are accompanied by very fine weather. These winds must not be expected here at any other time of the year. In the strong winds, generally varying between S.E. and S.W., but most frequently from the former quarter, it is dangerous for a ship to be caught near the land between Cape Howe and Wilson Promontory. On the Australian coast to the East of this strait, as well as on the eastern coast of Tasmania, northerly winds are sometimes found, veering to N.E., but they seldom attain any great strength.

Islands of Mauritius, Reunion, and Rodriguez.—These three islands are situated in the zone of the S.E. Trade winds, varying from E.N.E. to S.E., generally attended with fine weather, although sometimes they do drop rain.

At Mauritius the months of November and December are the hottest months in the year and those of the heaviest rains. In 1831, at Port Louis, the following observations were recorded of the winds:—In January and February the prevailing winds were S.E. and N.W.; in March N.W. and S.E.; in April S.E. and N.W.; the same in May; in June, July, and August S.E. regular; in September S.E. and N.W.; in October strong from S.E.; in November the winds are often variable; in December they are from East and S.E.

In the island of Reunion the months from November to April are the hottest and the period of heavy rains. On the shores of this island land and sea breezes are found, and often at the time of the syzygies southerly winds varying to West. During the fine season, from April to December, S.E. winds are constant, only varying from E.S.E. to S.S.E. The wind generally freshens about nine o'clock in the morning, and falls about four in the afternoon. During the night it is mostly calm, and if this is not the case, that it will blow hard the next day is pretty certain. During the day if the breeze is moderate and it ceases early, the land breeze continues during the night. In June, July, and August the breeze is generally very fresh. The S.E. wind prevails during the winter, from December to April, although it is generally weaker and often interrupted by calms and westerly and N.W. winds.

At Rodriguez the wind is generally from East and S.E., pretty fresh, with sometimes squalls and rain. These winds, are, however, generally accompanied with cloudy and foggy weather. Calms near this island are seldom found. It is from December to April, and chiefly near these islands, that those hurricanes are met with which render navigation so dangerous at this period in parts of the Indian Ocean.

Monsoons in the Molucca Channels.—In these seas two monsoons are distinguished, which seamen call the N.W. and the S.E. monsoons; some saying that the winds hang more to the northward than westward, and more to the southward than eastward. The first corresponds to the N.E. monsoon North of the equator, the second to the S.W. monsoon. It is known, indeed, that the monsoons which prevail in these channels are much less regular than in the open seas; and that according to the time of year the North and West winds prevail in turn, as well as those from South and East during the other monsoon. It may be noticed generally in these seas that South of the equator, as far as 10° or 12° South latitude, the direction of the wind varies 10 or 12 points from that of the prevailing wind North of the equator at the same time; that is, if a ship North of the equator have the wind from North, another ship South of it will have it from W.N.W., and if the first ship have the wind South the latter will have it from E.S.E. or East. But to avoid confusion arising from this the different names of the monsoons, the old names of the N.W. and S.E. will be here preserved, according to the case in question.

In the Java Sea, like that of the Moluccas, the N.W. monsoon commences in the first part of November; and does not attain its height till December. It continues till the end of March, a time when calms, light winds, squalls, rain, &c., occur.

The S.E. monsoon commences in April, gradually increasing till May: it ends in October, when the winds become variable.

Such is the general law observed in these two seas, but notwithstanding what has been said on the changing of the wind, it must be remembered that it sometimes draws to the northward or westward and sometimes southward or eastward. Besides this, the changes of the monsoons do not take place at regular periods. The S.E. monsoon is subject to calms, and the wind of it is not so strong as that from N.W.

Strait of Bally.—In the strait of Bally the wind often blows from North with much violence: in the strait of Sapy alternate land and sea breezes are found. They come from the South in the morning and from North at about two hours after noon, and are frequently separated by an interval of calm. In the other straits East of Java, a similar condition of the wind is found, and that also very variable.

North Coast of Java.—On the North coast of Java, from May to July, the wind is from S.E., with a return of opposite winds varying to N.E. Near the West point of the island during the S.E. monsoon, which brings the fine season, it is from S.S.E. varying to E.S.E. In October these winds become weaker and variable.

The N.W. monsoon generally begins in October, sometimes nearly a month before or after, and ends in March, being the season of the great rains. In December westerly winds prevail. Towards the middle of February there are storms and rain.

Batavia.—At Batavia from April to November the weather is tolerably fine; rains then set in for the rest of the year.

Borneo, South Coast.—On the South coast of the island of Borneo, that is from the Pulo-Laut to the strait of Sunda, the S.E. monsoon prevails from May to September, as it does on the West of Java. At this period, in the Indian Ocean, the S.W. monsoon prevails North of the line. From September to April the West wind blows on this coast, bringing constant rain and dirty weather. During the S.E. monsoon, the weather, which is still wet, is less rainy than during the N.W. monsoon. But it may be broadly stated that in Borneo it rains eleven months of the year.

Straits of Carimata, Gaspar, and Banca.—In the straits of Carimata, Gaspar, and Banca, when the N.E. monsoon prevails North of the equator, N.W. winds are mostly found.

North of Bouron and Ceram, the S.E. monsoon varies fram S.S.E. to S.S.W.; at Amboyna from East to S.E. In these islands the N.W. monsoon varies from W.S.W. to N.W. This last, often called the westerly monsoon, is the season of storms in these islands, ending in April. The other monsoon (the S.E.) begins in March and lasts till November; bringing the rainy season. In the Moluccas during this monsoon violent storms are met with; and rain falls, especially over the larger island of the Archipelago. In November this monsoon ceases. However, the northerly or N.W. monsoon does not become established for some time after, for during two months the winds are variable, as they always are in close seas towards the end of the monsoons. From October to April the weather is tolerably fine.

In the Moluccas, situated between 5° South and 1° North latitude, the winds are very irregular, as there is a great difference between the monsoons which prevail at the same periods in the two hemispheres.

West Coast of New Guinea.—On the West coast of New Guinea are two monsoons; one from the S.E. lasting from April to October; the other from N.W. beginning with the end of October and terminating towards April. In January, near this island the wind sometimes varies from N.N.W. to N.E. In the spring the weather is often changeable, and in March, April, and May the weather is squally. From June to September a great deal of rain falls; from October to May the weather is fine and calm, without either clouds or fogs.

Island of Celebes.—The island of Celebes, like that of Borneo, is divided by the equator into two parts, and the same phenomena are found here as described in referring to the monsoons at Borneo and Sumatra. On its South coast the S.E. monsoon is established from May to October, and the S.W. monsoon prevails at the same time on that part of the island which is North of the equator. The S.E. monsoon, lasting from May to October on the coast of Celebes South of the equator, brings the driest season. The N.W. monsoon replaces the S.E. towards October, and lasts till April; rain is then almost perpetual, and the wind strong.

During the two months when the sun is nearly over the island, and near the syzygies, we have always northerly winds and rain.

On that part of the island North of the equator, the N.E. monsoon in October takes the place of the S.W., making the fine season.

Strait of Macassar.—In the northern part of the strait of Macassar, from May to October, the S.E. monsoon is found on the East coast of Borneo. The same also takes place between Celebes aud Gilolo; it is succeeded by the N.W. monsoon, lasting from November to April.

In the southern part of the strait the wind is from the N.E. in April, May, and June, and there is less of it in August and September. During October, November, and December, as well as in the following months, in these latitudes fresh breezes prevail from W.S.W. to W.N.W.

Near the West coast of Celebes from May to October, land and sea breezes are found, while on the coast opposite of Borneo the wind is steady from South.

From November to April, on the West coast of Celebes, the wind varies from W.S.W. to W.N.W.; in April, May, and June it is from N.E., but light in the month of August.

It has been observed that when the S.W. wind prevails on the coast of Celebes, about six leagues off the coast it becomes W.N.W., and N.W. on the coast of Borneo. During the S.E. monsoon (from May to October) a vessel cannot work up against it on the low coast of Borneo, on which coast light land winds are found in this season; while on the corresponding coast of Celebes, which is elevated, a fresh land wind blows at night, followed by a sea breeze during the day. In December we generally meet the alternate winds near Celebes. In August and September the wind is light; but sometimes off this coast storms from S.W. occur and long calms.

Seas of Celebes and Sooloo.—In the Celebes Sea and Sooloo Archipelago easterly winds with fine weather prevail in October, but are not regularly established till November. In May the westerly winds replace them, and in a month become established to terminate in October; bringing with them a season made up of rain, squalls, and tempests, which take place principally in July and August. In September a heavy mist hangs about the coast of Mindanao.

At the beginning of the westerly monsoon the winds are light for some time, with heavy rain, during which the wind blows from an opposite direction, lasting from the eastward sometimes for above a week. Heavy storms occasionally happen until the westerly wind becomes established. During the whole of this monsoon the weather is cloudy, rainy, and sometimes stormy. In the same season, between Mindanao and Celebes, sudden and violent storms take place from N.W., the westerly winds sometimes lasting till November.

In the Sooloo Sea the East or N.E. monsoon is not a

steady fresh breeze but often variable. Near Mindanao the northerly winds never blow fresh, and light changeable winds often displace them for several days. The same occurs at the end of January, and it is considered that the same winds prevail from the Sooloo Archipelago to Manila.

Sea of Timor.—In the Timor Sea, and also in the Arrufura Sea between the Arroo Isles and the North coast of Australia, as well as in the vicinity of Torres Strait, the S.E. monsoon blows with much regularity. Towards the middle of it, from May to August, it varies from E.S.E. to S.E. and is then very strong. The Malays call this the white season. In the beginning and towards the end of the monsoon the wind is due East and sometimes veering to E.N.E. During this monsoon the wind is generally fresh and steady when the moon quarters, and we find calms and unsettled weather at the time of the syzygies. This fact has also been observed in the Trade wind of the eastern coast of Australia.

CHAPTER VI.

MONSOONS OF THE CHINA SEA, GULF OF SIAM, PHILIPPINE ISLANDS, AND NORTH-WEST SHORE OF BORNEO.

China Sea.—Two monsoons prevail in the China Sea, the South or S.W. monsoon and the opposite, North or N.E., and they succeed each other at intervals of six months. The S.W. monsoon blows from May to October; the N.E. from November to May, and they both have considerable influence on those of the adjacent seas. Like those of the Indian Ocean, these monsoons do not set in at the same time in all parts of the China Sea.

On the western coasts the S.W. monsoon begins in April, and is not established out at sea till a month after. It does not prevail on the North coast of Borneo, at Palawan, and the Philippines till between the middle and end of May. In the southern part of the China Sea it is found later than in the northern part of it.

In the Java and Molucca Seas the S.W. monsoon is not established till a month after it prevails in the China Sea. It lasts six months, and terminates as on the coasts of India, before it does out at sea.

The N.E. monsoon, which succeeds it, extends in the same manner successively to the southward out at sea;

so that to ascertain when the monsoon would commence in a particular part it would only be necessary to know when the opposite monsoon was established, and then allow it a duration of six months.

In the month of May light variable winds are often found in the China Sea, and with the S.W. monsoon, East or S.E. winds are occasionally found for a day or two. This occurs chiefly in the Northern part of the China Sea, and also frequently during either monsoon.

In the S.W. monsoon, especially from June to August, there are heavy rains, sometimes attended by violent squally weather. At the same time in the Formosa Channel the wind is frequently found varying between North and East. These three months are the strongest of the S.W. monsoon and those in which it is most steady. In September the wind is unsettled, often varying between N.E., East, and S.E. However, S.W. winds are still found to prevail, and last till the syzygy of the beginning of October. The monsoon then changes, and about this time a storm occurs, in which the wind comes from S.W., changing to West and N.W., bringing rain; the wind afterwards changes successively to N.N.E. and E.N.E., and then the N.E. monsoon is established. The month of October is a bad one for the navigation of the China Sea. Sometimes it is rainy; generally, however, at the close of the month the weather sets in fine.

The N.E. monsoon commences on the western coast of the China Sea about the 15th of October. In the southern part of this sea it is seldom found before the month of November. In some years September and

October are fine and the N.E. monsoon does nct invariably set in with a squall. It is at its height in December and January. This monsoon is sometimes attended with rain, and always produces much sea, especially near Pulo Sapata and between this island and Singapore. But it is not without its intervals of fine weather. The variable winds of October, November, and the beginning of December, which blow on the coast of Palawan, admit of sailing N.E. and S.W. during these months, but frequently with a cloudy sky and dull weather.

During the N.E. monsoon the wind is generally from North to N.E. on the coast of Luconia, and when it occasionally veers to N.W. or West it blows hard, with rain. This N.E. monsoon weakens in February. During this month and that of March it is but moderate, and the weather is then very fine in the China Sea.

Towards the end of the monsoons, when they lose strength, alternate land and sea breezes are found on the coasts. In March and April they are fresh and most regular. Towards the end of September and beginning of October they are also established, but are not so strong as in the former months, nor are they established so regular. At the change of the monsoons the wind is unsettled and variable. These periods are distinguished by stormy weather, and those of the September equinox are severe as well as those of October. In these months, as well as in November, the wind is fresh with occasional calms, but of short duration.

Coast of Malacca.—On the eastern coast of Malacca the weather is fine from April to October, while it is bad on the western coast of the same peninsula. On the eastern coast in the month of June a S.E. wind blows during the day; in the evening it changes to West, lasting till ten or eleven o'clock in the morning.

During the N.E. monsoon rain is constant on this coast, and the change from the S.W. monsoon to the N.E. is frequently attended by heavy storms. The return of the S.W. monsoon at the conclusion of the N.E. takes place calmly and quietly.

Gulf of Siam.—In the gulf of Siam the monsoons set in and terminate sooner than in the other parts of the China Sea, and the wind there is generally not so regular. The S.W. monsoon begins in April, followed by continual rain during May and June. In July, August, and September the weather is still rainy and the wind westerly.

On the coast of Siam, from March to May, southerly winds are often found, which from June to August become S.W., blowing very fresh, especially near Pulo Oby. From the month of September the wind becomes variable. In October the monsoon changes with a fresh breeze from S.W. In November, December, and January the weather is fine and wind steady at North. In February the wind varies between South and East, and during this month and the beginning of March the land and sea breezes are regular.

On the western coast of the gulf of Siam, in May, June, and July, when the S.W. monsoon is at its height, a land breeze prevails from three to twelve days together.

Coast of Cambodia.—On the coast of Cambodia, in June, July, and August, are heavy rains with S.W.

winds. On this coast the monsoons are not regular, and land and sea breezes are met with when the prevailing monsoon is weak. During the S.W. monsoon these breezes do not last more than five or six hours, and are not so fresh as those which prevail at the end of the N.E. monsoon.

Islands of Pulo Timoan and Pulo Condore.—In the islands of Pulo Timoan and Pulo Condore, the N.E. monsoon is established towards the 15th of October with fine weather. The S.W. monsoon brings rain and lasts for eight months. In November, near these islands, we have calms, storms, and rain, with tornadoes. At Pulo Condore the rains last for a month after the establishment of the N.E. monsoon. At Pulo Timoan the wind becomes unsettled in September and the change of monsoon brings bad weather. In November the weather is fine.

On the coast between the gulf of Siam and Cape Padaran the S.W. monsoon blows along the shore. Sometimes, near the land, during night, a light land breeze is found, which, falling calm, is succeeded by the wind of the monsoon, which blows fresh during the rest of the day.

On this coast the N.E. monsoon is established from the end of September or beginning of October till the middle of April.

Coast of Cochin China.—On the coast of Cochin China wintry weather is found with the cold northerly winds, accompanied by rain, that prevail from December to February. The season of the great rains includes the months of September, October, and November. During the N.E. monsoon easterly winds frequently occur. Between the Paracels and the coast the same winds prevail as far as Cape Varela; and in this channel calms frequently prevail while to seaward from this bank the monsoon is blowing fresh and regularly.

During the S.W. monsoon on this coast the land and sea breezes are tolerably regular, and the sea breeze is replaced every evening by a land breeze, which, although it does not always commence at the same time, blows every night, and is followed by a calm or a light wind, generally lasting till noon, when the S.E. wind again sets in.

On the coast of Cochin China the winds are variable during the whole year, and the monsoons mostly light. The leeward coast is not dangerous with the N.E. monsoon.

Gulf of Tonquin.—In the gulf of Tonquin we find two seasons, the wet and dry. The former is the hotter time of year, and begins towards the end of April, lasting until August, when the rains moderate, and September and October bring tolerable weather. In November commences the dry season, and about the middle of this month strong northerly winds set in, varying to Eastand E.S.E. towards the close of it.

Towards the end of December the wind from N.N.E. becomes East, accompanied by fog. During January and February it is from N.E. and N.N.E. In April the wind is not so fresh, nor is the weather so cold. In the gulf of Tonquin typhoons are met with.

Coast of China.—On the coast of China, from the island of Hainan to Amoy, S.W. and N.E. monsoons are subject to the same variations generally as those of the China Sea. But on the coast opposite to Formosa it is said that northerly winds prevail for eight or nine months. During the N.E. monsoon on the South coast of China the wind is E.N.E. along the land, sometimes drawing along the coast, and at others to the S.E. We rarely, however, find regular land and sea breezes; those we speak of are more solar breezes. On the South coast of China, during the S.W. monsoon, southerly and S.E. winds are frequently met with. In June, July, and August there is a great deal of rain there and the weather is dark and cloudy.

On the East coast of China, North of Canton and as far as the Chusan Islands, as already observed, the two monsoons prevail.

The N.E. monsoon on the coast comprised between Macao and Chusan generally begins in October and lasts till the beginning of May. This monsoon is at its height from November to January: it is generally a double-reef breeze. There being always a troublesome sea with these winds, which are steady between N.E. and North, ships cannot carry much sail to them, and should always keep as close to the shore as possible, as the sea there is smoother than outside.

In the N.E. monsoon heavy winds, lasting for two or three days, will set in, which are considered to belong to a typhoon at sea. When this violent wind ceases, for several days a light breeze sets in from the East. But after a week or ten days it again freshens, blowing hard, and producing a heavy sea, especially at the entrance of the Formosa Channel.

In October, November, and December, the weather is

tolerably fine, and not a day passes without a break in the rainy and cloudy weather. From January to March a great deal of cloud and fog hangs about. In April the N.E. monsoon loses strength for a day or two, sometimes the wind changes to South, when the weather then becomes thick, and heavy squalls follow from N.E. till the middle of May.

The S.W. monsoon, varying to S.S.W. and South, begins in June or July, and ending in October, is attended with storms and typhoons.

In the zone of these monsoons the rainy season on all the coasts facing the West is during the S.W. monsoon; and on the contrary this is the dry season on all those coasts facing the East. When the N.E. monsoon prevails in its turn, the coasts facing East are subject to rain, while those facing West have dry weather.

Formosa Channel.—In the Formosa Channel, between that island and China, bad weather may be looked for at all times of the year. Even in the summer, and also towards the middle of the S.W. monsoon, stormy and squally weather prevails, accompanied by torrents of rain.

In the channel comprised between Formosa and the Philippines, bad weather is found at all seasons. In the N.E. monsoon the wind is principally from E.N.E. as far as the opening of the channel, where it becomes N.b.E. and N.N.E. About the Bashees it is N.E. and E.N.E., causing a heavy sea. On the eastern coast of Formosa variable easterly winds are found, favourable for ships going northward of the island, but which become there N.N.E. As a general rule, it may be considered that where the wind hauls gradually to the southward of East, it soon returns northward again with double its strength. Heavy N.W. winds sometimes last for several days following on the North coast of China, and reach to a good distance out to sea. The barometer is of little use in giving warning of these winds, common as they are, for it is very high with a strong northerly wind which sometimes prevails. When the wind has gone round the compass, it generally returns suddenly to North, blowing so hard as to compel ships to shorten sail.

Philippines.—Among the Philippine Islands prevail the two regular monsoons which are met with in the China Sea. They even sometimes extend as far South as the Marianne Islands in the Pacific Ocean, and as far North as the coast of Japan.

The Philippines being a group lying North and South, their high land naturally interrupts the regular course of the wind: and hence it is, that at forty or fifty leagues from them ships encounter so much bad weather sometimes, and which becomes worse as the islands are approached.

The N.E. monsoon begins about October, with fine weather, lasting till April, and varying to the northward; but if it should veer occasionally to N.W. it blows hard.

The S.W. monsoon is not known here till between the beginning and end of May: nor does it become regular till June. During this monsoon the weather is gloomy, cloudy, without a tint of rain. Sometimes about this period severe storms occur, called *Collas Tempestados.*

They are generally accompanied by thunder and rain, the wind changing about and blowing with the same force from all points of the compass. These collas and bad weather take place at the end of July or middle of August, and sometimes in October. They are not unlike the typhoons.

In September the wind loses strength, the rain is less, and the sky becomes fine; but this in some degree is compensated by morning fogs, which last till noon.

At the change of the monsoons bad weather is sometimes experienced as in the China Sea.

Island of Luconia.—In February and March, about the end of the N.E. monsoon, on the coast of Luconia, the wind is variable, and often with a tendency to follow the course of the alternate land and sea or solar breezes. In the month of April, the alternate winds are well established; and from June to October, the period of the S.W. monsoon, the wind which blows on the coast at right angles to it brings rain.

Island of Palawan.—The island of Palawan does not present anything in particular, and the monsoons near it are the same as those of the China Sea. Near the West part of this island, however, in September and part of October, there are strong S.W. winds, accompanied by dark and rainy weather. On the North coast in these months and in December, the wind is variable, but fair for vessels from N.E. or S.W.; but the weather is generally cloudy at this season.

Island of Borneo.—The island of Borneo forms the N.W. and western boundary of the China Sea. This island is intersected by the equator; and, as we find in

Sumatra, the consequence is that the monsoons of the N.W. coast do not happen at the same time as those on the West coast. Thus the S.W. monsoon prevails on the N.W. coast from May to October, while at the same period the S.E. monsoon prevails on the West coast; and the N.E. monsoon blows on the N.W. coast while the N.W. monsoon prevails on the West coast.

The S.W. monsoon is not established on the northern part of Borneo till between the 15th and 30th of May; the rain then continual. In September the weather is not so bad, and the dry season sets in with the N.E. winds, varying to East. However, this can hardly be considered the dry season, for, owing to its position under the equator, the island is inundated with rain.

On the western coast the S.E. monsoon prevails towards the end of May, bringing fine weather. From September to April the West or N.W. monsoon prevails, with heavy rain, and sometimes strong gales.

Typhoons.—The typhoons of the Philippines and the China Sea sometimes extend as far westward as Calcutta. They are not always preceded by any signs which might place the mariner on his guard against them; in some instances they are preceded by fine weather, with a clear sky and fresh or moderate breezes, generally blowing in an opposite direction to the prevailing monsoon, which is S.W., for the typhoons only happen between June and November. The prelude to one of them is frequently a thick heavy dark cloud near the horizon in the N.E. The upper edges of it are copper-coloured, and the higher they are the more vivid is the colour. When this cloud rises and moves rapidly, the typhoon bursts, producing rain, thunder, and lightning; sometimes an hour of calm succeeds the storm, which generally lasts twelve hours; then the S.W. wind blows almost as hard, and establishes the equilibrium. The barometer is the best indicator of these tempests. Near the coast of China they generally begin from N.W. to North; the wind afterwards veers to N.E. and East, from which quarter they blow strongly, causing a heavy sea and strong currents setting westward. The wind then changes from East to S.E., then to South, after which they become more moderate. The shifting of the wind in this direction during the typhoon, is only found in those which sweep along the coast of China, for at the distance of forty or sixty leagues from the land, these changes of the wind are experienced in a contrary direction:* that is, the wind beginning at North or N.W., instead of veering successively to N.E. and East, changes to N.W. and West, whence it blows violently, afterwards decreasing in proportion as they reach S.W. and South.

Typhoons rarely take place more than once in an interval of three or four years. The zone in which they blow extends from 30° to 10° North latitude, although they rarely pass 16° North latitude. With regard to longitude, they occupy the space comprised between the coast of China and the meridian of 150° East.

* This results from being on the opposite side of the line or course on which the whole meteor is advancing. The reader should refer to the "Storm Compass" already alluded to.

CHAPTER VII.

A GENERAL VIEW OF THE CURRENTS OF THE INDIAN OCEAN AND CHINA SEA.

In the Indian Ocean the equatorial current is found only between the parallels of 10° and 25° S. lat.; that is, only in the zone occupied by the S.E. Trade wind. Commencing near the meridian of 100° E. long., it reaches the African coast with some few interruptions, which will be noticed hereafter. On the above meridian the counter-current of the Indian Ocean meets it, to which our attention shall be first devoted.

Counter-Current of the Indian Ocean.—This current which is also the counter-current of the Cape of Good Hope, is formed of two others. One is a branch of the counter-current of the Atlantic; the other is a branch of the Agulhas current, returning into the Indian Ocean a portion of the waters which run S.W. and South along the East coast of Africa. In fact, this counter-current of the Indian Ocean, running easterly, mingles its waters with those from the South Polar regions flowing towards the equator: the result of which is, that its first direction becomes modified, and it sets E.N.E. and N.E.

Off Cape Leuwin, the S.W. point of Australia, this

current is divided into two branches. One, flowing northward along the West coast of Australia, mingles with the equatorial current about the tropic of Capricorn. The other, called the Australian current, extending from the coast as far as the parallel of 40° S. lat., or thereabouts, flows eastward and E.N.E. along the South coast of that continent. South of this current variable currents are generally found.

The temperature of this current on the parallel of 40° has been found about 56°; on that of 30°, from 66° to 70°. This last temperature was between the meridians of 92° and 102° E. long.

The rate of this current is very variable; on the surface it depends on the winds, concerning which we are deficient of information. Off Cape Agulhas the mean velocity of it is about thirty miles a day. Its breadth is often 210 to 240 miles, and it is mostly found between the parallels of 37° S. lat. and 40° or 42°. It increases in proportion as it advances eastward, and it would be difficult, from the nature of its formation, to say what are its precise limits.

This current is particularly important to vessels passing from the Atlantic to the Indian Ocean, and crossing it from the westward for the Pacific. From the island of St. Paul to Cape Leuwin, ships have been set 200 miles within those limits; and it appears in October, November, and December, the current increases in rapidity as the season advances.

Australian Current.—The branch of this current called the Australian current, after flowing along the South coast of that continent, gradually diminishes as it progresses eastward, and does not run into Bass Strait, where variable tides and currents are found, generally depending on the wind. It turns to the southward of Tasmania, and its southern limit is in about lat. 45° S. It then enters the Pacific and joins the counter-current of this sea, to which it contributes all it can.

Off Cape Leuwin, to the westward, the temperature of this current varies from 65° to 61°. It then gradually decreases as it advances eastward, and on the meridian of Tasmania it is between 47° and 49°, which is its lowest temperature. It afterwards increases as the current enters the Pacific.

The velocity of the Australian current is very variable. Near Cape Leuwin it has been estimated at twenty-eight and twenty-nine miles a day. As it advances eastward, towards Bass Strait, it decreases, and varies between sixteen and thirteen miles.

Equatorial Current.—The equatorial current, as we have already said, is only found South of the equator in the Indian Ocean. Its northern limit is about the same as the equatorial limit of the S.E. Trades, namely in about 10° S. lat. On the meridian of 100° E. long., its southern limit often reaches as far as 25° S.: but this limit, as the current advances westward, inclines northwardly, and decreases the breadth of the current, which, on the meridian of the island of Rodriguez, does not extend beyond the parallel of 20° S. lat.

On the meridian of Rodriguez it separates into two branches: one, flowing S.W., surrounds the island of Mauritius, and its northern limit passes South of Isle

Réunion* and afterwards South of Madagascar: on which meridian this branch is 480 miles across, and has a temperature of 75° and 76°. From this meridian its breadth decreases rapidly, and, flowing to the S.W., it joins the Agulhas current South of Natal. At their meeting the breadth of the equatorial current is not more than about seventy or eighty miles, and its temperature has been estimated at 70° or 72°.

In that space comprised between the Agulhas current, the S.W. branch of the Equatorial current, and the western coast of Madagascar, the currents are very variable, sometimes setting E.S.E. and S.E. near the northern limit of this current, and North and N.E. near the eastern limit of the Agulhas current. The rapidity of these counter-currents is often considerable, and sometimes attains fifteen or twenty miles in twenty-four hours.

As already observed the S.W. branch of the Equatorial current flows round the islands of Rodriguez and Mauritius, but does not reach Isle Reunion; near which, to the northward and eastward, currents are sometimes met with setting N.W., N.N.W., and North. Off this island the velocity of the S.W. current varies from ten to fifteen miles a day. Off the South point of Madagascar it has been found to be from forty to fifty miles.

The West branch of the Equatorial current, which may be regarded as a continuation of this current, runs W.N.W. and, passing round the northern end of Madagascar, enters the Mozambique Channel, joining that

* Late Bourbon.

current and flowing with it along the East coast of Africa. Off Cape Amber, its velocity has been estimated at from forty-five to sixty miles a day, and its temperature has been found about 78°. This branch of the Equatorial current is about 360 miles across. After joining the Mozambique current the Equatorial current forfeits its name, and pursuing their course along the East coast of Africa, on reaching the latitude of Natal these two obtain the name of the Agulhas current.

This great current flows along the South coast of Africa, and on gaining the meridian of 20° E. the main stream forms into two. The first mingles with the counter-current of the Indian Ocean, already noticed, and the second, flowing into the Atlantic Ocean, continues N.N.W. and N.W. along the West coast of Africa, joining the southern portion of the Equatorial current of the Atlantic.

The velocity of the Equatorial current is very variable; nevertheless it attains a rate of twenty to twenty-two miles a day. Its mean force may be estimated at twelve or sixteen miles a day.

The temperature of this current, from its eastern commencement as far as the meridian of Rodriguez gradually increases, and there varies from 72° to 82°.

Mozambique Current.—During the whole year, and particularly about the equator, a constant current sets along the East coast of Africa from N.E. to S.W. towards the Mozambique channel, from which it derives its name, attaining considerable strength, particularly off Cape Corrientes.

Arrived at Cape Padrao, the Mozambique current

here changes its name, taking that of Agulhas or the Cape current. Off this cape it receives the S.W. branch of the Equatorial current, and flows with greater velocity near the coast than it does at a distance from it.

The mean velocity of the Mozambique current varies between eighteen and twenty-eight miles a day. Off Cape Corrientes it has been found to be forty-six leagues in the twenty-four hours. There is no other instance of so strong a current, unless perhaps it be in the full strength of the Gulf Stream. The Mozambique current is stronger during the S.E. monsoon than in that of the N.W.; and is accounted for by the accumulation of water forced by the wind on the whole African coast of this sea. It varies much in breadth, according to the different parts of the coast.

On the West coast of Madagascar a counter-current is found setting northward, said to be partly occasioned by the Mozambique current.

Off Cape Guardafui, in the parallel of this cape, the temperature of the Mozambique current has been found at 86°. As the current progresses southward the temperature falls. In the Mozambique channel it is not more than 65° or 66°; but, receiving there the Equatorial stream, heated as it has been by the powerful sun, about Natal the temperature becomes 82° or 84°.

Agulhas Current.—The peculiarities of this current were first studied in 1777 by Major Rennel. It is particularly strong in the winter, and still more so on the bank, and even outside of it, where it is strongest. But near the coast, on the contrary, it is very weak, a fact the knowledge of which is important to vessels entering the Indian Ocean from the Atlantic to profit by in keeping the shore on board; while, on the contrary, a vessel running westward would skirt the edge of the bank. Although this current endures through all seasons and at all times, it is yet controlled by circumstances. Thus, a gale from West or S.W. will check its velocity, and sometimes overcome it altogether. In such cases it returns on its course with redoubled force as soon as the gale is over. At other times it continues in spite of the gale, producing of course a terrific sea. A vessel in this case would hug the land; where she would find less wind, a smoother sea, and a current by no means so strong, which conditions are very important in voyages from the Atlantic to the Indian Ocean, and *vice versâ*.

The strength of the Agulhas current varies considerably. Sometimes it is only fifteen miles a day; at others, seventy, eighty, and even 120 miles a day.

In June, July, and August, between the meridians of 40° and 37°, and the parallels of 30° and 30½° S., the current between the coast of Africa and Madagascar generally sets W.S.W. as far as the eastern edge of the Agulhas Bank.

Off the coast of Natal it takes a S.W. direction along the land, until it reaches the edge of the bank between the meridians of 30½° and 30° E. It there acquires more strength and flows nearly W.b.S. and W.S.W. along the edge of the bank, as far as the meridian of 23° E. On this meridian the edge of the bank lies S.W., and afterwards S.S.W. ½ W. to its southern extremity. The current there also changes, flowing S.W. as far as the meridian of 22° E.; from whence it flows

S.W.b.S. to its southern extreme. There the current is seldom strong beyond the parallel of 36½° S., and the meridian of 20°. Beyond this a slight current continues flowing westward, losing itself in the Atlantic Ocean. Afterwards the strength of the current flows round the edge of the bank to the N.W. as far as the Cape. But this N.W. current seldom attains half the velocity of that which flows S.W. along the S.E. edge of the bank.

A counter-current setting eastward, formed of the waters of the same current of the Atlantic Ocean, is often found at the southern limit of the Agulhas current. It is most frequently met with between the parallels of 36½° and 40° S.

Variable Currents North of the Equator.—The above are the only currents which may be considered general currents of the Indian Ocean. North of the equator, in this sea, in the Molucca Channels, and in the China Sea, the periodical winds, alternate breezes, and changes of monsoon, combine to produce a great variety of currents. The direction of these currents depends on that of the winds which have produced them, on the form and character of the coasts where they are met, and on many other conditions. It would be beyond our limits to go into their details; but facts connected with the more important of those coasts may be stated, where a certain regularity in character and periodical return may be observed. And although, in respect of the rest, they might be studied with advantage, we shall merely allude to the works which treat more minutely of the navigation of the Indian Ocean.

Arabian Gulf.—North of the equator, in the Arabian Gulf, the currents take the direction of the wind. They are seldom strong in this extensive sea, except in August and September, when the large rivers falling into it produce, with their increased waters, a southerly current, varying from eight to thirty miles a day. This current is especially found on the parallel of 20° N. In October, when the N.E. monsoon prevails, it takes a more southerly direction. In November it increases with the monsoon, and its strength out at sea varies from eight to thirteen miles a day; it runs W.S.W. and S.W. along the coasts of Arabia and Africa between eight and thirty miles. In January it is little felt at sea; in March it is variable out at sea; and in April it runs to N.E., varying from N.N.E. to E.N.E., with a velocity similar to that during the preceding monsoon.

Currents of the Red Sea.—The currents of the Red Sea appear to be entirely regulated by the prevailing wind, submitting always to its influence. When a southerly wind prevails, the current flows northward; and when succeeded by a northerly wind, sets southward. Its force depends equally on that of the wind, increasing or decreasing in the same proportion, and during the variables, which precede the periodical winds, scarcely any current is observed.

Off the port of Jeddah, in different seasons, the current attains a velocity of a mile or a mile and a half an hour. North of Jeddah it varies much during the whole year, entirely governed by the direction of the wind; which, when it is strong, impels it at a rate of twenty, or even forty, miles per day.

From May to October, in the northern part of the Red Sea, the level of the water is 1·9 feet lower than during the other months of the year, the effect of the strong northerly winds which prevail during that season over the whole extent of the sea. They produce a continual current, flowing from the strait of Babel Mandeb into the gulf of Aden. This current is so rapid that, even with a stiff breeze from S.W., a vessel sometimes cannot make head against it.

From October to May, when southerly winds prevail in the lower part of the Red Sea, the current changes its direction, and sets rapidly to the northward. The mass of water, finding no outlet, accumulates in the northern part of this sea, attaining a greater elevation than at any other period.

The temperature of the water of the Red Sea has been found at 86° in the strait of Babel Mandeb; 85° off Mocha; 83° to 81° South of Jeddah; 77° in the North part of the Red Sea; and 70° at the extremity of it.

On the coast of Arabia, from September to March, in the N.E. monsoon, the current sets strongly to the westward. In March and April it runs S.W.; in June, July, and August, it turns eastward. About Aden, from April to August, it flows N.N.E. and East; and from this latter month till March it runs E.S.E. and S.E.b.E.

Persian Gulf.—From May to September a current flows into the Persian Gulf, and from September to May it sets outwards. In March and April, during the little shemaal, the current sets northward, so that a vessel working to windward may make good twenty or twenty-five miles in twenty-four hours.

At Ras Jask, from September to April, the current sets southward; and the same is found in the space between that part of the coast and the opposite coast of Arabia. This current extends as far as the parallel of 23° N., but it is very seldom at this period that the current is found setting towards the Persian Gulf.

From March to September, while northerly winds prevail in the gulf, the current in the strait leading to it sets against the wind with so much strength, that a vessel might make twenty miles a day to windward.

From October to March vessels entering the Persian Gulf, or leaving it, should keep near the Persian coast. From March to September it is more advantageous to keep mid-channel or on the Arabian coast.

In October, on the coast of Persia, with the N.E. monsoon, the current flows southward till March; in April it changes with the S.W. monsoon, and flows northward during the six following months. When, however, the S.W. monsoon is at its height, the current sets rapidly to the eastward along the Persian coast.

The Indus.—The Indus, the mouth of which is situated in the N.E. angle of the Arabian Gulf, discharges a considerable volume of water into this sea, especially from July to September. At this period it produces a current setting to the southward at first, and which, spreading as it gains distance, extends its western and eastern borders to S.S.W. and S.E. The current of this river, which is very narrow at its source, attains a breadth of 240 miles after a course of 300 miles; its strength gradually decreasing as it flows seaward. At that distance it is but small, and the current becomes lost in the ocean.

West Coast of Hindostan.—In November the currents are generally weak and variable on the West coast of Hindostan, except near Anjenga and from thence to Cape Comorin, where strong S.E. currents are often found. On this same portion of the coast, between Anjenga and Cape Comorin, a northerly current prevails during January.

From November to March, the period of the fine season (N.E. monsoon), a current is rarely found on this coast.

In March and April, when N.W. winds are strong, they produce a small south-easterly set along the coast: otherwise no current exists.

In May a southerly current flows along the whole West coast of Hindostan at the rate of five to seven miles a day, increasing near Anjenga and Cape Comorin. Sometimes, however, in May and June it is scarcely perceptible, and sometimes there is none. In this season, when the wind veers to the southward, the current takes a northerly direction. This always occurs on the northern part of the coast, near Bombay. In the same months the current out at sea generally sets E.N.E.

In July, when the rivers are increased by rain, the current sets rapidly to the southward along the coast.

Nevertheless, at the entrance of Bombay, we rarely find a southerly current; that occasioned by the discharge from the rivers generally runs to the westward, and is brought back by the tide. South of Bombay, the current sets southward at the rate of ten or fifteen miles a day, and sometimes as much as eighteen or twenty miles. These currents, which are also found during August and September, are more rapid between Cochin and Cape Comorin, varying to the middle of October from twenty to thirty miles per day. When the rivers overflow (in June and July) the current generally sets off the coast, though sometimes it sets directly towards it.

Between Calicut and Anjenga, in the end of September and beginning of November, a S.E. or E.S.E. current is found setting at the rate of a mile or a mile and a half per hour.

Laccadives.—About the Laccadives, during the S.W. monsoon, on approaching these islands southerly and S.W. currents are found, particularly in March. Between the islands their direction is S.S.W. In the channel separating them from the Malabar coast they set about S.S.E.

In the Nine Degrees channel the current generally runs S.W., and rarely to the northward. In February, March, and April its direction is S.S.E.

The rate of the current during the S.W. monsoon between the Laccadives and the coast varies from twenty to twenty-six miles a day. Amongst the islands it is from eighteen to twenty-two miles a day.

From forty to sixty leagues westward, in the offing, the currents flow West and W.S.W. at the rate of eight to eleven miles a day.

Everywhere between the Laccadives and the coast of Malabar the current flows westerly, and near these islands it sets to the southward during the S.W. monsoon at a rate varying from fifteen to thirty miles a day. Nevertheless, between the islands sometimes a northerly current at the rate of eight or ten miles a day is found.

Maldives.—It has been observed that about the Maldives the current generally sets more to the eastward than in any other direction: this, however, is modified by local causes.

When the S.W. monsoon prevails North of the equator, that is, during June, July, and August, the current sets easterly in the vicinity of the North Atoll, while at the same time generally in July and August, when the S.E. Trade wind blows South of the equator, the current sets W.N.W. near the South part of the group in the equatorial channel and about the South Atoll.

To the eastward of the North Atoll, in March and April it generally runs West, and, on the contrary, near the South Atoll, it sets E.N.E. between the equator and the parallel of 4° or 5° South latitude. This current often flows at the rate of fifty and sixty-five miles a day; but sometimes it is feeble and variable. In the same months, between the equator and 8° or 9° North latitude, the current sets S.W. In May it runs strong to the Eastward near the equator, and sometimes attains a rate varying from fifty to seventy miles a day near the South Atoll, between the parallels of 2° N. and 2° S. latitude.

In the Equatorial channel and in the latitude of the South Atoll, from October to January the current flows eastward; while it flows westward with great force at the same time between the parallels of 3° and 5° N. lat., and between the Maldives and Ceylon. In the islands

North of the equator at this period the current runs S.W., and in those situated South of this line it runs South and S.S.E.

Thus it is evident that for the most part the currents are governed by the monsoons from N.E. and N.W. as well as those from S.W. and S.E., which prevail at the same period to the North and South of these islands.

During the N.E. monsoon, on the meridian of Point de Galle, and especially in November and December, the current is found setting at the rate of thirty and thirty-five miles; but decreasing in rapidity as it nears the equator.

Gulf of Manar.—Throughout the gulf of Manar the currents are uncertain during the N.E. monsoon; sometimes, however, they have been found to run to the S.W. at a rate of eighteen or twenty miles. From May to September, during the S.W. monsoon, a current runs into the gulf at the rate of eight miles a day; running to the northward it follows the direction of the coasts, and sets into the bay of Palk, to make its exit between the points Pedro and Calymere.

Ceylon.—On the East coast of Ceylon, during the N.E. monsoon, from the middle of September to February, especially in October and November, a strong current sets to the southward along the coast; rounding the South coast it assumes a westerly direction as far as Point de Galle, and sometimes as far as Colombo. Its mean rate has been estimated from forty to forty-eight miles a day; it has been sometimes found to amount to ninety-four miles, under certain circumstances; but at others it is weak.

During the same N.E. monsoon, on the western coast of Ceylon the currents are very variable. Sometimes a fresh N.N.E. wind blowing for a day or two produces a tolerably strong current setting northward along the coast. It would appear that the current in this case is only a continuation of that which on the East coast takes a southerly direction, and on the South coast of the island flowing westward as just observed, and which near Point de Galle again makes towards the West coast.

During the S.W. monsoon the current we have been considering turns, and on the South and East coasts takes directions nearly opposite to those abovementioned; that is, it flows eastward on the South coast of Ceylon, and northward on the eastern coast. Its rapidity is then by no means so great as during the other monsoon, at least near the coasts of this island, and it decreases in proportion as it is near the shore. Off the N.E. part of the island, however, it has been estimated at from ten to twenty miles a day.

From May to September no current is found on the eastern coast of the island; while out at sea it sets strongly to N.N.E. and N.N.W. When the westerly winds are in full force, Nichelson says that the currents cease near the land and for a distance of fifteen or eighteen miles from it.

Gulf of Bengal.—In the gulf of Bengal the current depends entirely on the monsoons. During the N.E. monsoon it flows southward and S.W., and during the S.W. monsoon it flows northward.

In April the general movement of the waters to the

North and N.E. commences over the whole surface of the gulf, and continues till the beginning or middle of October. In April and May the current attains its greatest force, its rate varying from ten to twenty miles a day.

In the middle of the gulf, during March and April, there is but little current, and what little there is is variable; between the Coromandel coast and the Nicobars, as well as at the entrance of the strait of Malacca, it frequently sets S.W. In the northern part of the gulf of Bengal the current often flows southerly in March; but in April it oftener flows northerly.

During the S.W. monsoon, in the eastern part of the gulf, and near the strait of Malacca, a current is often found setting to the southward, like that near the Malabar coast.

At the distance of forty or fifty leagues from the coasts, in the middle of the gulf of Bengal, the S.W. current begins regularly about the middle of October or beginning of November, about the time when the N.E. monsoon is well established. Before this period the current is variable; we find it setting S.S.W. and even N.W. or somewhere between these two directions.

In January this current loses strength, and in February it entirely ceases, in the middle of the gulf and perhaps on the coasts where if it still exists it runs northward and N.W.

West Coast of the Gulf of Bengal.—On the coast of Coromandel, as on the coasts of Golconda and Orixa, the current begins flowing northward about the end of January or beginning of February; its rate increasing in April and May, when the South wind prevails with great regularity. After May it fails, and ceases according to the monsoon in August or September. Sometimes it is met with as late as the middle of October, and during the other months it generally runs southerly. The northerly direction of the current along the coast of Coromandel and all the eastern coast of Hindostan, from Ceylon to Balasore, varies between North and N.E., and sometimes even E.N.E. In August the current from the Ganges is strong, and extends as far as the coast of Coromandel, contributing to modify the current flowing northward, and also decreasing it. During the S.W. monsoon the rate of the current varies between a mile and a mile and a half an hour. Towards the middle of October the current assumes a southerly direction along the Coromandel coast. In November and December it attains the rate of one mile an hour, setting S.S.W. From the beginning of December to January it is influenced by the wind, and runs southward at the rate of two or two and a half miles an hour. In March it sometimes takes a northerly direction, but most frequently flows southward during this month.

East Coast of the Gulf of Bengal.—On the eastern shore of the gulf of Bengal, and near the strait of Malacca during the S.W. monsoon, the current sometimes sets to the southward as above stated.

During the N.E. monsoon, near the Nicobar Islands, and also between these islands and that of Junkseylon, a strong N.W. current is found, and sometimes setting North. On the coast between Cape Negrais and the bay of Chittagong, in both monsoons, the current is seldom strong, and sometimes tides are experienced. However, from the middle of December to May a southerly current is sometimes found there; but from June to September it runs northward.

Throughout the bay of Bengal, in the month of December, there are no currents but those produced by the tide that prevail during the greater part of the year. From September to November, however, on the coast of Bengal, a current is often found setting between W.N.W. and W.S.W.

West Coast of Sumatra.—On the West coast of Sumatra the current is generally weak, not more than a half or three quarters of a mile an hour.

It is stated that from October to May it follows the S.E. monsoon, and the N.W. from May to October. Forest, however, shows, that during the S.W. monsoon the mass of water driven towards the West coast of Sumatra is divided on reaching Cape Acheen into two branches, one flowing to N.E., the other to S.E. along the coast. It is rarely found running North on this coast of the island, but most frequently S.E.

After the month of August, near Acheen, and along the S.W. coast of Sumatra, as well as on the South coast of Java, the current sets strongly to the eastward. At Acheen, after the month of December, it sets from the strait of Malacca to the westward, while near the S.W. coast of Sumatra it sets southward.

Sunda Isles.—The currents are very rapid in the narrow channels between the Sunda Islands. They generally follow the direction of the prevailing winds, and are grealy influenced by the tides. Consequently they cannot be described with any certainty. In the strait of Sunda, however, from January to April, the current flows to the eastward, and during the rest of the year it takes a westerly direction, sometimes running at the rate of three miles and a half an hour. Between the Sunda Isles and the coast of New Holland the current sets to the eastward from November to April, while the N.W. monsoon lasts, and to the westward during the S.E. monsoon, from April to November.

CHAPTER VIII.

THE CURRENTS OF THE CHINA SEA.

During the N.E. monsoon the current in the China Sea generally runs S.W. with a velocity depending on the strength of the wind. At the commencement of the monsoon, when it is strong, the S.W. current increases; and when on the contrary the monsoon becomes weaker, there is little or no current.

Western Part of China Sea.—In the western part of the China Sea, along the Malacca and Cochin China coasts, a southerly current generally begins about the middle of October. On the Cochin China coast it is sometimes earlier and continues till the month of April In March it constantly runs southward near Pulo-Aor, a season when calms and light East winds often prevail in those latitudes.

On the same coast of Cochin China and near the island of Hainan, the current, varying from South to S.W., commences about the middle of September; near the coast from 15° to 11° N. lat. it becomes much stronger; but loses in strength in proportion as it flows southward.

During the N.E. monsoon, from the parallel of 14° N. lat. to Cape Padaran, the current near the coast runs

South at the rate of forty, fifty, and even sixty miles in twenty-four hours. The rate, however, is variable, and it is only in the zone abovementioned that it is occasionally so strong. It is weaker near Cape Padaran, as well as at the mouth of the gulf of Siam, and the direction of it is then S.W.

South Coast of China.—Near the South coast of China, at this period, the current generally assumes a W.S.W. direction, and nearly parallel to the coast. It is sometimes very rapid after strong winds or typhoons; but out at sea, at about twenty-five leagues from the land, it is not so strong as it is inshore and between the islands. Sometimes near the land the westerly current is overcome by the tide.

Formosa Channel.—In the Formosa Channel, between this island and the coast, the current sets southward during the N.E. monsoon; and when the N.E. winds are strong, it takes a S.W. and southerly direction in the strait between the South end of Formosa and the North point of Luconia. When, however, the wind is variable, the current often sets northerly.

West Coast of Luconia.—On the western coast of Luconia the current is very variable and flows sometimes southerly and sometimes northerly.

Coast of Palawan.—The same occurs on the coast of Palawan, where it depends entirely on the wind, and only becomes strong in gales.

In the middle of the China Sea a northerly current is found in the beginning of May, and sometimes in the end of April. When the S.W. monsoon prevails it sets northerly and N.E. until September. During this monsoon, however, and when the wind is light or moderate, the current is not so steady, and takes different directions. After the strength of the monsoon is past, a N.E. current is seldom found at sea: but sometimes instead of it one running South.

Gulf of Siam.—Along the coast of Cambodia, from Pulo Oby to Cape Padaran, the current generally runs N.E. parallel with the coast from April to October. At the same time, off the East coast of Malacca, from the strait of Singapore to the gulf of Siam, the direction of it is N.W.

Coast of Cochin China.—Off the coast of Cochin China, North of Cape Padaran, the current is not so strong during the S.W. monsoon, and when found between this cape and the gulf of Tonquin it is setting alternately to the northward and southward.

When fresh winds from N.W. or West prevail in the gulf of Tonquin, as well as to seaward, the current near the Paracels and in that part of the sea where these winds extend, generally sets S.W. or South; and its direction in this case being oblique or nearly opposed to that of the wind, it produces a very dangerous sea.

South Coast of China.—On the South coast of China, as in other parts, the wind has great influence on the current. During the strong S.W. winds it sets eastward parallel to the coast. It is generally strongest near the islands West of Macao, and between these islands there is frequently a westerly current, produced by the waters of the Canton River. Its rate between Macao and the island of St. John is from one to two miles an hour, and its direction varies between W.S.W. and W.N.W.

This current is not experienced during the S.W. monsoon; sometimes it decreases, and then a weak tide is found flowing eastward.

Coasts of Luconia and Palawan.—On the coasts of Luconia and Palawan, the current sets to the northward during the whole of the S.W. monsoon; but it is sometimes scarcely felt, and even near the coast is seldom strong.

Bashee Isles.—When strong westerly winds prevail near the Bashee Isles, an easterly current is found. Most frequently, however, it runs strong to the northward, varying at times from N.N.W. to N.E.

Molucca Channels.—In the Molucca Channels the current during the N.W. monsoon is much stronger than during the S.E., which accounts for our being able to cross the strait of Sunda from the Indian Ocean more easily in May, June, July, and August than when leaving those channels in January and February. Besides which, in those seas the current depends in a great measure on the winds, notwithstanding their variable character.

CHAPTER IX.

NAVIGATION:

OUTWARD ROUTES ACROSS THE INDIAN OCEAN.

THE rapid progress of navigation has been no where so remarkable as in the Indian Ocean, and in no other sea has it produced results so important. Not only is the seaman now informed whereabouts in this sea, according to the seasons, he will find those periodical winds and currents that are favourable to his purpose, but he may take his opportunities for regulating his voyages and profiting by them accordingly; and this knowledge is not only useful to navigation in general, but is especially important to vessels without steam.

In reference to similar demands of the seaman in the Atlantic Ocean, the routes to be taken to the Cape of Good Hope have been shown. We shall now consider those to be adopted from thence to ports in the Indian Ocean and China Sea, availing ourselves of what M. D'Après de Mannevillette and Horsburgh have said on the subject.

From the Cape of Good Hope the mariner has the choice of several routes to different parts of the Indian Ocean. That which he selects must depend on the *season* of the year, and therefore on the monsoon that is then blowing.

To facilitate his choice, against each of the passages are arranged the several periods when they should be adopted, as follows:

ROUTES.	WHEN TO BE ADOPTED.
No. 1.—*The Inner or Mozambique Route.* No. 2.—*Route outside or East of Madagascar.*	During the S.W. monsoon, when certain of reaching the port of destination before it is over, being those of the Red Sea, the Persian Gulf, and ports of India or the straits of Malacca,—*see p.* 111, 113.
No. 3.—*Boscawen Route.*	Adopted when bound to the western ports of India at an advanced period of the season, when the port might not be gained before the N.E. monsoon commences,—*see p.* 115.
No. 4.—*Middle Route.*	Is principally taken when the Cape is passed during the month of September by vessels bound to the western coasts of India,—*see p.* 115.
No. 5.—*First Outer Route.*	Adopted for the gulf of Bengal, but with the certainty of arriving before the end of the S.W. monsoon,—*see p.* 119.

No. 6.—*Second Outer Route.*	Adopted when it is likely the gulf of Bengal will not be reached before the S.W. monsoon is over, or after the N.E. monsoon has set in,—*see p.* 120.
No. 7.—*Direct Route for Sunda Strait.*	Adopted during the S.W. monsoon, and when certain of gaining the port of destination before this monsoon is over,—*see p.* 124.
No. 8.—*The Eastern Routes.*	Taken when the meridian of St. Paul and Amsterdam is passed between the middle of September and October, so as to reach the port of destination with the N.E. monsoon. In this route one of the eastern straits, Bally, Lombock, Allass, Sapy, or Ombay, is adopted,—*see p.* 128.

These are the eight principal routes which are generally adopted in crossing the Indian Ocean from the Cape of Good Hope outwards, and here numbered for easier reference. They will equally serve for homeward voyages to the Cape, and will be followed by directions for the navigation of the Arabian Gulf as well as the gulf of Bengal, and

Lastly, by a discussion of the routes to be taken in the China Sea, and also the inland seas, and the routes from Australia to the Chinese ports, as well as those of the Indian Ocean.

No. 1.—*The Inner or Mozambique Route.*

The route by the Mozambique Channel is the most direct to India during the S.W. monsoon, from April to November, for vessels bound to Bombay, Ceylon, and the gulf of Bengal; but it should be adopted only when they are sure of reaching the Indian coast while it lasts.

In making for the Mozambique Channel a vessel should avoid the S.W. and westerly currents found after leaving the Agulhas Bank; and with this view should not make to the northward too soon, more especially if it be desired to sight the coast of Madagascar southward of St. Augustine Bay. It would generally be as well to gain the meridian of 37° E. before crossing the parallel of 34° or 35° South latitude, and then make directly for the channel.

A vessel may then proceed along the coast of Madagascar or take the middle of the channel. By adopting the former course she would have to guard against dangers off the coast; and in the latter, which appears to be mostly preferred, to keep clear of the Bassas de India and the Europa Rocks. In crossing their latitudes at night, a vessel would keep either to the East or West of them, and having passed them should steer for Mohilla and Comoro, leaving the island of Juan de Nova to the eastward.

With a fair southerly or S.W. wind, Horsburgh says, pass to the westward of the Bassas de India and the Europa Rocks rather than between them and Madagascar, because the African coast is cleaner than the other.

But on approaching that coast light winds and a southerly current are found, and the monsoon is usually stronger in the middle of the channel than on either shore. However, in April, and in the beginning of May, the best wind is found West of Comoro, a little to the West of mid-channel.

From Comoro, when bound to India, a vessel should cross the line in 54° East, and when North of it may steer for her port of destination. Thus, if bound for Surat or Bombay, the route from the equator is direct; only keeping more to the West on the parallel of the high land of St. John for the first, and on that of the isle of Kundary for the second port, and then make straight for them.

Vessels bound to Goa should make the land at the Burnt Isles; and those for ports southward of the western part of India, should pass through the Eight Degrees or Nine Degr es Channel. Nevertheless when the port of destination is Cananore, Tellichery, Mahé, or Calicut, the route may be shortened by passing North of the Laccadives, from whence she may make for Pigeon Island, and then run down the coast to the southward.

When bound to Cochin, Quilon, or even Anjengas, the Nine Degrees or the Eight Degrees Channel, is generally preferred, the Degree and a Half Channel might also be adopted. This latter channel is preferred by vessels bound for Ceylon and the Coromandel coast. A vessel having passed through it and sure of her position, or if she should have sighted Minisoy, having gained to the eastward of this island, may steer direct for Point de Galle. A vessel uncertain of her position should run to the eastward until she strike soundings on the bank off Cape Comorin, between the parallels of 8° and 9° North latitude; and if the weather is clear, which rarely happens during the S.W. monsoon, land will be seen at the distance of twenty-seven or twenty-eight miles. As soon as soundings are obtained, a course to the southward should be adopted, in order to reach the latitude of Point de Galle, for with the S.W. winds the current sets toward the gulf of Manar, and off Point de Galle the wind frequently becomes S.S.W., delaying vessels considerably if they cannot weather it.

By the first route Krusenstern sailed from the Cape on the 27th of June, 1797, crossed the line in 44° 30′ E. long., and reached Ceylon on the 2nd of August, and Madras on the 4th, after a voyage of forty days.

No. 2.—*Route East of Madagascar.*

The route East of Madagascar is often preferred, because perhaps it is less dangerous than the former, and is certainly less liable to variable winds; hence it might occur, that two ships separating to the eastward of the Cape of Good Hope, one of them to take the Mozambique route, and the other that East of Madagascar, that the latter vessel might be the first to reach her port of destination. This last is often adopted in the months of August and September, the time when variable winds are found in the Mozambique Channel.

The route East of Madagascar is adopted from February to October. Nevertheless it is necessary then to have doubled the Cape at a time so as to be sure of reaching the port of destination in India with the S.W. monsoon.

When taking this route, a vessel from the Cape should steer eastward, and take care not to get so far to the northward as to cross the southern limit of the trade winds before she has reached the meridian of 50° East. By meeting the Trade wind too soon, much difficulty will be experienced in making easting, particularly on the parallel of the S.E. part of Madagascar, where winds, known by the name of Fort Dauphin breezes, generally blow strong from E.N.E., producing a current, (sometimes from forty to fifty miles,) setting to the southward near the shore. These winds are found several degrees East of Madagascar.

In crossing the trade, a vessel should keep between the meridians of 51° and 52°, until she reaches the parallel of 15° S.; she would then sight the eastern cape of Madagascar, or the land northward of it, or perhaps Cape Amber, thence taking a fresh departure. In leaving Cape Amber the best course is North or N.b.E., until the islands situated to the N.E. and N.W. of this cape are passed. Then with the S.W. monsoon she would easily reach any port on the coast of India.

A vessel leaving the Cape for the Red Sea or Persian Gulf from September to March, will find it most advantageous to adopt this second route East of Madagascar.

From March to September she should take the Mozambique Channel. Nevertheless this second route might still be adopted, although the first is shorter; but having passed the Seychelles, the Red Sea or Persian Gulf should be steered for direct.

No. 3.—*Boscawen Route.*

The Boscawen route is called after the Admiral whose name it bears; who, commanding a fleet composed of twenty-six sailing vessels, adopted it on his way to India, and made a very short passage.

In this route the Admiral passed to the South of Madagascar, keeping between the parallels of 37° and 38° South, and in the counter-current of the Indian Ocean, until he had arrived in 54° or 55° East. He then steered northwards and passed between the islands of Reunion and Mauritius. Leaving these islands he steered directly North, with the islands of Cargados Garajos, and the Sayha de Malha shoals to the eastward, and Galega and the Seychelles to the West, crossing the line on the meridian of 62° East.

The Boscawen route is generally taken by vessels bound to Bombay or the Malabar coast at such an advanced period as to render their reaching it before the commencement of the N.E. monsoon uncertain. For vessels which have not doubled the Cape before the 1st of September, this passage and the middle one, alluded to presently, are more advantageous than the two preceding them. Indeed, these two routes being further eastward, a vessel would be in a better position to reach her port should she meet with the N.E. monsoon.

No. 4.—*Middle Route.*

The Middle route, which is still more easterly than the Boscawen, leads outside and to the East of Madagascar, and then West of the Chagos Archipelago. A vessel taking this route from the Cape, should pass East of the meridian of 67½° or 68° East, so as not to find the Trade wind in the case of their hauling more easterly, which ordinarily happens in March or April. Thence making less northing she would cross the parallels of 27° or 26° South latitude. On reaching the Trade she would cross it steering North, keeping on the meridian of 67½° or 68½° as long as the wind will permit.

Near the equator westerly and N.W. winds are frequently met, and she would profit by them as much as possible in making northing. Beyond the equator the N.E. monsoon might not be a foul wind, because at this period it is oftener N.W. than N.E. If not too strong it is favourable to vessels bound to the Malabar coast.

When the Trade is crossed by a vessel going northward, she should keep between the meridians of 65° and 68°, in order not to be set down to the Maldives by the easterly current and the light winds met with near the equator. In October and November, when in 3° or 4° N. latitude, a vessel is very likely to meet with northerly winds, inclining more westerly than easterly, as before observed. She would profit by these slants of wind, taking the most advantageous tack; as the sea is generally smooth, in making these tacks well she would easily pass to the North of the Maldives, in which case she would run to the N.E. for the land. If instead of being bound to a northern part of the West coast of India, she were going to a southern port of that coast, she would take the Nine Degrees Channel or the Eight Degrees Channel between the Maldives and the Laccadives, or the Degree and a Half Channel if she be bound to Ceylon.

The Middle route is taken mostly by vessels which double the Cape of Good Hope in the month of September, when bound for the western coast of India.

A vessel going from the Cape to the Mauritius, should make her easting in a high latitude, according to the foregoing instructions for the Middle and Boscawen routes, continuing more or less to the eastward according to the season. When she meets the Trade she should be on the meridian of the island to which she is bound, or to the eastward of it, in order that she may not be obliged by the wind being at East to keep too close a luff in order to lay up for the island.

But when the sun has high North declination, it is not necessary for a good sailing ship to stand so far East as above said, for then the Trade is between E.S.E. and S.E., the time when it has least Easting. In March and April, on the contrary, it is frequently due East; and from November to March it is oftener to the northward of East than due East, varying indeed from N.E. to N.W.

From March to September vessels leaving Reunion or Mauritius for India, should proceed northward passing East or West of the Cargados Garajos, and East of the Seychelles. Then if bound for the West coast of India, after they have crossed the line they should adopt one of the routes, according to their port of destination.

If bound to the gulf of Bengal, after having passed the Seychelles they should take the Equatorial Channel or the Degree and a Half Channel, and make for Ceylon. On leaving the Mauritius or Reunion for India or the gulf of Bengal, it will be more advantageous, if the wind permits, to pass East of the Cargados Garajos as well as the Chagos.

During the N.E. monsoon, in leaving the above islands, Horsburgh recommends a vessel going to the gulf of Bengal to adopt the same route as in the S.W. monsoon; and he considers that a vessel should stand far enough to the eastward to take the Equatorial Channel or that of the Degree and a Half, this route being more direct than the other.

D'Après, on the contrary, says the route to be preferred is that taken by him, and which he calls the equatorial route. He shows that several vessels by following this route have reached Pondicherry from Mauritius in twenty-six or thirty days.

A vessel leaving Reunion or Mauritius in either of the months of November, December, January, February, and the beginning of March, should stand to the northward or take the Boscawen route until she has reached the parallel of 5° S., when she should make to the eastward; and, according to D'Après, keep between the parallels of 4° 40′ and 3° South latitude. Horsburgh says, she may even keep between 3° and 1° S.; but as the former says to the northward of the parallel of 2° S. is a district of calms and squalls, while between 3° and 4° S. there are often light westerly winds. If she is bound to the gulf of Bengal, in standing to the eastward when she has reached the meridian of 90° or 92° East, she should make to the northward to cross the line, and may then steer for her port of destination;

having then to make less easting if bound to a port on the West coast of India, and in which case reaching the meridian of 81° or 82° East would be sufficient.

The four routes abovementioned may be taken not only by vessels bound to the western coasts of India, but also by those for the gulf of Bengal, provided that the Cape is left in sufficient time for them to reach their ports of destination before the end of the S.W. monsoon.

No. 5.—*First Outer Route.*

With the same conditions a vessel may make a more direct passage to the gulf of Bengal from March to October, and proceed as follows by the first of the outer routes.

From the Cape of Good Hope she would keep nearly on the parallel of 38° or 39° S., as far as the meridian of 62° E. From thence steering about N.N.E. she would cross the southern limit of the Trade (the parallel of 28° or 29°) on the meridian of 82° or 83°. Then continuing northward, she would take care in crossing the Trade to gain easting, gradually increasing her longitude in order to be prepared for change, (as in the months of March, April, and May the wind will be found frequently at East,) so as to compensate for the effects of westerly currents found in those latitudes. Thus a vessel in 1° or 2° South latitude, between April and October, may be certain of reaching any of the ports in the gulf of Bengal with the S.W. monsoon.

Vessels bound for Ceylon or Madras, should stand to the northward across the Trade, and keep a little West of Point de Galle if desirous to make it. If bound to Trincomalee in this season, they should make the land a little South of their port. Those bound to Pondicherry or Madras should also, from March to September, make the land South of those ports, to provide for the effects of the northerly current produced by the S.W. monsoon.

No. 6.—*Second Outer Route.*

A vessel bound to the gulf of Bengal, if we suppose that from the Cape she would cross the Line between October and April, and not arrive in the gulf until after the N.E. monsoon has set in, or when it is in its greatest force, should adopt the following route, called the second outer route.

From the Cape she should run sufficiently to the eastward to cross the southern limit of the Trades on the meridian of 84° or 85° E., and would then steer to the northward and gradually gain Easting sufficient to counteract the effects of the current, so as to enable her to lay up for Cape Acheen, the northern point of Sumatra.

She should not, however, go so far East as to sight the islands off the West coast of Sumatra or Cape Acheen, for in October and November light variable winds and squalls from N.W. or West prevail there, with a current setting into the strait of Malacca. If she be bound to this strait, on the contrary, this route would be the best to follow.

When she has passed Cape Acheen, at fifty or sixty

leagues West of it, if the wind permit she would sight the western part of the Nicobar Islands. When the wind appears likely to veer to the eastward, she would keep as much to windward as possible; and if it veers to E.N.E. or N.E., she would pass West of those islands. On the parallel of 16° or 17° the wind frequently comes from North, and she would profit by it in making some boards to the East, in order to keep off the West coast of the gulf. But she should not approach too near the coast of Arracan, as it is generally in the northern part of the bay of Bengal that off these coasts it is necessary to work to windward, and thus reach the anchorages of Bengal and Calcutta.

A vessel bound to Pondicherry or Madras or any other port on the West coast of the gulf, need not approach Cape Acheen as abovementioned, and should not go East of the meridian of 87° E. In fact, in October and the beginning of November a vessel off that cape is often delayed by N.W. or westerly winds; while during the same months, in the middle of the gulf, the wind varies between South and West.

After the month of September a vessel should make the land a little North of her port of destination, to counteract the effect of the southerly current. This, however, depends on the changes of the monsoons, which are very far from being regular, and sometimes vary a whole month. Sometimes the monsoon is as much before its time as it is sometimes as much after.

It has been long ascertained that in the route just mentioned it is by no means necessary to pass to the eastward of the Nicobar and Andaman Islands; this route on the contrary is unfavourable on account of the light winds and southerly currents found near them during the N.E. monsoon.

Vessels leaving the straits of Malacca for the western ports of the gulf of Bengal, Pondicherry, or Calcutta, do not take the East coast; they adopt the channel between the Andaman Islands, passing between the Great and Little Andaman, and sometimes even to the southward of the Nicobar Islands.

No. 7.—*Direct Route for Sunda Strait.*

For a vessel bound through the strait of Sunda, if she does not touch at the Cape, the best route is to enter the Indian Ocean between the parallels of 37° and 38° S., and thus avoid the current across the Agulhas Bank; and she will then take the counter current of the Indian Ocean, favourable to her.

From the meridian of the Cape she would keep between the parallels of 36° and 40°, to sight the islands of Amsterdam and St. Paul. However, many seamen prefer the route between the parallels of 36° and 37°, where the wind they say is fresher and more constant than in higher latitudes.

On leaving the meridian of Amsterdam and St. Paul, she would steer so as to cross that of 80° E. in 38° S., and the meridian of 90° E. in 33° S., steering then so as to cross the tropic of Capricorn in about 102° E.

In making to the eastward after leaving St. Paul, should she meet with contrary winds she should not pass to the northward of 30°, because in a lower latitude easterly and N.E. winds are often met. It will be more advantageous to keep to the southward in order to profit by the westerly winds on the parallel of 38°.

From March to September, and especially in March, April, and May, a vessel should hasten to reach the meridian of Java Head, and then stand to the northward, because the Trade often veers to the eastward, and the westerly current is strong near the South coast of Java. It is important, therefore, to make the land East of the strait of Sunda at this period, and also East of Bencoolen if bound to this port.

From May to the beginning of July, a ship bound to Bencoolen should not sight Java Head, for the passage is more direct at this period, and she should make for the island of Engaño, and then for Bencoolen, if the wind permits. In these months the wind often varies to N.W., and then the current sets S.E.

From September to March, between the equator and the North limit of the Trade, the N.W. monsoon is found. This monsoon produces an easterly current, and frequently extends as far as the parallel of 14° S.

At this season a vessel bound through the strait of Sunda should not steer North when on the meridian of Java Head. She should make for the S.W. extremity of Sumatra, or the island of Engaño, and pass westward of this cape. When she is on the parallel of Sunda Strait, she should make for it, endeavouring to compensate for the effect of the southerly currents which prevail at this period. It is important to observe these directions from November to the beginning of February.

At the same period, when bound for Bencoolen, a vessel on losing the S.E. Trade, should keep on the meridian of Cape Acheen until she is well to the northward of Keeling or Cocos Island, or until she has reached the latitude of Java Head. She will then probably find N.W. winds to take her to the island of Trieste; passing North of it if she pleases, and South of Isle Larga, when the wind has a tendency to veer northerly. If it veers to S.W. or the southward, she may steer direct for Bencoolen, passing South of Trieste.

The foregoing directions may be completed with some observations on the strait of Sunda that may be useful to vessels crossing this strait for the Java Sea.

The strait of Sunda is divided into several channels by islands. The southern one, called the Princes Channel, was formerly one of the most frequented; but with a fresh breeze the channel of Crockatoa is generally preferred, lying between the island of that name and Princes Island, or between Crockatoa and Pulo-Bessy.

Princes Channel.—On entering the Princes Channel with the N.W. monsoon, a vessel should keep near this island and the Charpentiers Rocks; during the S.E. monsoon it will be better, especially with a good breeze, to keep near Java and Frere Island. When she has reached the second point (the first is Java Head) she should make direct for the fourth, and pass it at a convenient distance. When abreast of Anjer, she should steer so as to pass between the Cape and Bouton, at a convenient distance from each, taking care to keep well off the Brouwers Bank. After passing this bank and Bouton, she should steer for the Deux Frères if going to the strait of Banca, and pass two or three miles from Point Bantam if she is bound to that bay or to Batavia.

Crockatoa Channel.—When a vessel adopts the Crockatoa Channel, she should pass North of Princes Island, giving it a good berth; from thence she should steer for the fourth point as abovementioned.

Pulo-Bessy Channel.—The channel of Pulo-Bessy affords the advantage of anchorage, so that it is sometimes preferable to the Crockatoa Channel, more particularly when working out of it with the westerly monsoon. The Hindostan Rock is the only danger known in it.

Middle Channel.—Vessels homeward bound from China, with the N.W. monsoon, when leaving the strait of Banca, often adopt the channel between Middle Island and Sumatra. This channel is one of the most direct.

North Channel.—When the monsoon is strong, the North Channel may still be adopted, because it can be rapidly passed with the assistance of the westerly current. It should only be adopted when with a fresh breeze.

No. 8.—*The Eastern Routes.*

The Eastern route, through one of the Eastern straits, is taken about the end of the season, that is, when leaving the meridian of the islands of Amsterdam and St. Paul from the middle of September to February.

The general route adopted is as follows. A vessel first proceeds according to route No. 7, making St. Paul and Amsterdam. From thence she proceeds N.E. so as to cross the meridian of 111° E., in about 30° S., and then stands to the Northward, gradually nearing the Australian coast, till she is on the parallel of its N.W. cape.

Some seamen consider that this cape should be sighted, but this is by no means necessary; yet it should be passed at a distance depending on the season and the prevailing winds, and especially according to the strait which is to be adopted.

The prevailing winds on the western coast of Australia, between Cape Leuwin and the N.W. cape, are from S.W. and South. From April to November the S.E. or easterly monsoon prevails near the southern coasts of the Sunda Isles; at which period these winds may be expected everywhere between these islands and the North coast of Australia. But from November to March, when the N.W. or westerly monsoon prevails near these isles, variable winds are found there, although they generally veer to the westward; and as the current follows the monsoon, a vessel on leaving the parallel of the N.W. cape should keep to the westward of the strait she intends to take during the N.W. monsoon; and, on the contrary, during the other monsoon she should rather keep eastward of it. A vessel anticipating westerly winds when making for the straits of Bally, Lombock, or Allass should pass the N.W. cape of Australia at a convenient distance, and steer directly for her adopted strait.

If the S.E. wind fails her between the parallels of 18° or 14° S. lat., and she finds westerly winds there, she should be careful not to get to leeward of the strait she is making for, as the wind is often W.N.W. and even N.W. on the South coast of the Sunda Isles, between Java Head and the island of Timor.

To the foregoing observations may be added some on taking the channels for China by the eastern routes.

Strait of Lombock.—The strait of Lombock may be easily known by the large island of Banditte which divides its entrance into two nearly equal channels. The eastern channel is that generally used, and a vessel taking the western one must avoid the West point of Banditte and give the island a good berth in light winds.

A vessel taking the Lombock Strait should keep in mid-channel between Banditte and Lombock, and afterwards nearer to the eastern than the western coast. The current in this strait is very rapid, and no soundings can be had.

Allass Strait.—A vessel taking the Allass Strait,—which is safe and much frequented,—should keep nearer to the Lombock than the Sumbawa shore, because she can get soundings near the S.E. point of the former, while the latter shore is very steep.

Strait of Sapy.—The strait of Sapy is also much frequented, but should not be taken during the N.W. monsoon, at which time the strait of Allass is to be preferred. A vessel making for the strait of Sapy with light variable easterly winds should sight the western extremity of Sandal-wood Island. With westerly winds, which are generally strong, she should near the South coast of Sumbawa. A vessel once in the channel may pass through it to the sea of Java, passing either East or West of Gilibanta.

Straits of Flores, Alloo, and Pantar.—The straits of Flores, Alloo, and Pantar, between Ombay and the islands of Flores or Mangerye, are generally not much frequented. They are small, with strong currents, and therefore Ombay is preferred, or perhaps one of those West of Flores.

Ombay Channel.—A vessel approaching the N.W. coast of Australia for one of the channels into the Pacific Ocean, will find the Ombay Channel the most preferable. She should sight the eastern point of Sandal-wood Island; then pass between it and Savu, or between Savu and Rotte, if she is baffled with contrary winds from N.W. or West. Besides, it is more advantageous, especially with moderate winds, to pass East of Sandal-wood Island than West of it and in the channel which it forms with Flores. In case of having strong easterly currents with the N.W. monsoon, she would, on the contrary, take the route to the West of Sandal-wood Island, and pass between that island and Flores.

HOMEWARD ROUTES ACROSS THE INDIAN OCEAN.

Return from the Red Sea, Persian Gulf, or Ports on the West Coast of India, by Route No. 1.

Vessels bound from the Red Sea, Persian Gulf, or any of the ports on the western coast of India, in the course of the N.E. monsoon, to the Cape, may adopt the route No. 1 by the Mozambique Channel. It is the most direct route for leaving the Indian Ocean in this monsoon from the Red Sea or Persian Gulf; but from any of the ports on the western coast of India it should not be taken during the height of the monsoon, in December and January, when the N.E. winds generally prevail throughout the Mozambique Channel, or at least reach very nearly to the southern part of it.

It is important in all cases when leaving the western coasts of India, not to adopt this route either too soon or too late; that is to say, neither towards the beginning nor towards the end of the monsoon. Notwithstanding it is the shortest for leaving the Indian Ocean, it might happen that a vessel would be much detained by southerly winds frequent in October and November as well as in February and March.

The strong southerly current flowing is undoubtedly favourable, nevertheless this route should not be taken when leaving the West coasts of India with the N.E. monsoon. Besides, in January and February heavy squalls have to be encountered in the southern part of the channel, and for this reason most seamen always prefer the route No. 4 (the Middle Passage) when leaving the western ports of India.

From the Red Sea or the Persian Gulf the route is direct, and when near the African coast a vessel should follow these directions and those subsequently given for passing the Cape. First steering for the Comoro Isles, and then passing North of the Seychelles and the islands in the N.W. part of the archipelago of Madagascar, a vessel should pass West of Comoro or take either of the channels between this island and Mayotte. If the wind be fresh from the northward she may keep in mid-channel; if, on the contrary, the wind be variable or southerly, she should near the African coast in order to profit by the favourable currents which prevail there. She should always pass to the West of Juan de Nova, the Europa Rock, the Bassas da India, and make Cape Corrientes if the weather admits, then preserve a distance from twelve to thirty miles from the Natal coast, unless the S.E. wind brings a heavy sea, in which case she would keep further from it.

On Passing the Cape of Good Hope.

In returning from India it will always be best to make the land somewhere about Algoa Bay, especially should it not have been seen to the northward of it. The coast may then be kept at a moderate distance, and the Agulhas Bank may be adopted, the edge of which should not be passed. In case of sighting the coast to the eastward of Cape Recif or East of Algoa Bay in foggy weather, and a vessel is obliged to stand in shore working to westward, it will be necessary, particularly at night, to keep a good look-out.

In February, March, and beginning of April, a vessel having sighted the land about Algoa Bay, the S.E. winds being prevalent at this time, the best route for her to follow is the outer edge of the Agulhas Bank to the meridian of 23° or 24° E.; and in order to counteract the effects of the current setting W.S.W., S.W., and S.S.W. in this part, she should keep rather near the land so as not to get off the bank.

During the winter months, (June, July, and August,) when westerly and N.W. winds prevail, a vessel should keep near the land, as the sea is smoother there than in the offing. D'Après de Mannevillette says that it should not be left further than thirty-six miles, and he gives this as the proper distance. He adds, "As the soundings on this bank vary very much, and as the coast is very steep in several places, sounding is not a sure means of ascertaining the distance from it, and therefore great attention should be paid to the navigation. When coming from sea, the land is often concealed by fog, which forms a curtain over it, hiding the low part of it from view, while the summits of the mountains are easily seen."

At all times of the year when the wind is strong from N.E., and varying by the North to West, it is always best to sight the coast, and to steer so as not to get to the southward, for a vessel might find herself in the counter-current of the Indian Ocean, setting her easterly.

In general there is no danger in nearing the African coast, for the wind seldom blows hard directly on it, and a vessel can mostly stand off it on one tack.

Sometimes to the East of the Cape bad weather is found: while West of it, after doubling it with great difficulty, fine weather is found, and strong southerly breezes.

When a ship is on the meridian of Cape Agulhas, with the wind fresh from S.W. or S.S.W., she should stand well out so as to double the Cape well to the southward, a precaution particularly necessary during the night.

Should circumstances prevent a ship from sighting either the Cape or Cape Hanglip, she should at least get soundings on the western part of the Agulhas Bank, to be sure that she is far enough West. This precaution is indispensable in order to enable her to shape her next course. In fact, when a vessel is not certain of her longitude, she should keep constantly sounding, and should not steer so as to double the Cape until she has lost soundings on the western edge of the bank. In the event of a ship having shaped her course N.W. and again finding soundings, she should immediately alter her course to West. The western edge of the bank runs South by East from the Cape. As to the southern edge of it, this is rather composed of several little separate banks than by the continuation of the same bank. It often happens that in lat. 36° S. the bottom is not reached even to the eastward of the Cape. The soundings are therefore in such a case not to be trusted; but sufficient precaution having been adopted to be sure that the ship has passed it, a N.W. course may be shaped.

Return Route from the West Coast of India and Ceylon.

Vessels from the western coast of India, Ceylon, or Pondicherry, bound for the Cape of Good Hope, should follow the route above indicated. From the West coast of India with the N.E. monsoon they cannot cross the equator so far East as those from the gulf of Bengal, the consequence of which is, that they are often carried near Rodriguez by the S.E. Trade winds.

Leaving the West coast of India, a vessel should keep along the coast as far as the S.W. coast of Ceylon and Cape Dondra. From this cape she should stand S.E. with the N.E. monsoon, by which she may cross the line between the meridians of 84° and 85° E.

From April to November stormy weather is very rarely met with in the Indian Ocean; and a vessel should then pass thirty or forty leagues East of Rodriguez. During the other months it is better to give it a berth of seventy or eighty leagues.

From Mauritius or Reunion for the Cape.

On leaving these islands as soon as the ship is to the southward of them she should steer so as to pass at a distance of about thirty leagues from the S.E. part of Madagascar. From the parallel of 26° 30′ or 27° she should steer W.S.W. until the African coast is made about Algoa Bay, or she may even make the coast as far up as Port Natal, to profit by the current which flows along it.

Besides the advantages of this being a direct route, it admits of a vessel remaining as long as possible in the region of the Trade winds without any ill consequences should she happen to meet with a different wind in this part. As soon as she has sighted the land she may follow the instructions given above for doubling the Cape.

Vessels leaving Madras or Pondicherry with the N.E. monsoon, to pass Ceylon whether sighting it or not should cross the line between 86° or 87° or still further East in 89°, and then adopt the most convenient course for Mauritius or Reunion, if desiring to anchor there, or pass to the southward of them not caring to do so; from thence she would follow the route abovementioned.

From Bengal or the Straits of Malacca to the Cape.

Vessels from Bengal or the straits of Malacca bound to the Cape, should cross the line in 89° or 91° East. Leaving the strait of Malacca it is best to stand well out on a W.S.W. course before making to the southward for the purpose of avoiding the light variable winds of the islands on the S.W. coast of Sumatra. In case of finding light winds near the equator, they should take that course which makes most southing, in order to reach the S.E. Trades as soon as possible. As soon as they have found these winds they may shape a course so as to pass at a convenient distance from Rodriguez, Mauritius, or

Reunion, if not desirous of anchoring there, and then follow the routes above directed.

From the Straits of Sunda and the Eastern Straits to the Cape.

Vessels leaving the strait of Sunda or those to the eastward for the Cape of Good Hope, should navigate so as to reach as soon as possible the zone of the Trades, and also that part where they are the strongest, and then make westing. The Trade winds are generally found fresh in 14° and 15°, and stronger in 18° and 20° S. latitude. They are generally strongest between these two last parallels, although in March and April they are often light and sometimes even interrupted by westerly winds.

CHAPTER X.

NAVIGATION OF THE ARABIAN GULF AND RED SEA.

From Bombay or Malabar Coast to the Red Sea.

The most favourable season for the passage from ports on the western coasts of India to the Red Sea, is from October to April, and especially so is the interval between the 1st of February and the middle of March.

Vessels leaving Bombay after April should take the southern route, which we shall point out hereafter, and make their westing as they near the equator. At this period they meet with strong southerly winds near the African coast in the neighbourhood of Cape Orfui or Ros-Hafoon, during the S.W. monsoon. In working up from Cape Guardafui to Burnt Island, they should have good sails and strong rigging, for they will meet with boisterous weather.

During the month of May the wind is not so strong and has a more southerly direction than during the other months, and this in the gulf of Aden enables them to run easily to the West along the African coast. In May a vessel may make her passage to Aden with greater certainty than at a more advanced period of the season.

Leaving Bombay or any of the ports situated on the northern part of the Malabar coast for the Red Sea in November and December, a vessel should steer so as to pass between Socotra and the Arabian coast and then make West for Aden. The strictest look-out should be kept, and frequent soundings obtained. At this period the wind is fresh, especially West of Socotra; the weather is cloudy and the land difficult to make out.

In January and February nearly the same winds prevail, but they are more moderate and the sky is generally clearer. At this season the N.E. point of Socotra, if necessary, can be readily made, and then the North coast of this island passed. After this a course may be shaped for Aden; or the vessel may pass North of the island without seeking to make it, as before observed, and then steer for Aden.

In March and April the wind is not so steady as in the four preceding months (November to February). It often varies from N.N.W. to N.N.E., and calms, light breezes, and sometimes squalls, alternately follow each other. However, the weather is generally fair during one month. On leaving Bombay at this season, a vessel should pass southward of Socotra; for after the beginning of April the N.E. monsoon ceases near this island, as well as near the coast of Arabia, and instead of it light breezes from S.W. to West and frequent calms take place. Northerly currents also generally prevail at this season North of Socotra, and also between this island and Cape Guardafui. It is therefore advisable towards the end of March or beginning of April to pass about twelve leagues South of Socotra, so as to make Cape

Guardafui with the wind from S.W., which ought to be found there.

A vessel later in the season and having made the land, would find it prudent to stand well in for Cape Guardafui. Should the month of April be far advanced she should also keep the land on board as far as Burnt Island, and then make for Cape Aden. From May to August, when strong S.W. and W.S.W. winds prevail, it is sometimes very difficult while in shore to get from Cape Guardafui to Burnt Island. It is, nevertheless, the best course to follow, and it will be prudent not to steer for Aden till Burnt Island is passed. A good sailing ship may reach the strait of Babel-Mandeb during the height of the S.W. monsoon by working to windward, taking care to profit by every favourable circumstance. At new and full moon westerly currents will be found, and also slight variations in the winds. At this season also short passages may often be made by hugging the African coast until about sixty miles West of Burnt Island; from whence a course may be shaped for the strait of Babel-Mandeb, should the wind permit.

From Southern Ports on the Malabar Coast to the Red Sea.

Vessels from Anjenga, Cochin, Calicut, Mahe, or other ports situated on the South coast of Malabar, should run to the westward during the months of November, December, January, and February, and take that channel of the Laccadives which is the most direct for them. A vessel leaving Anjenga or Cochin should pass southward of Seuhelipar, keeping on the parallel of 9° 20′ or 9° 30′ N. lat.; while from Mangalore or Mahe she would pass North of all the islands.

In March and April, when North and N.W. winds are yet found between the Malabar coast and Africa, it is better to follow the Indian shore as far as something North of Mont-Dilly, and then shape a course to the northward of all the islands and banks.

Ships leaving Anjenga or Cochin, should take the Nine Degrees Channel and approach Kalpeni and Seuhelipar Islands, to allow for currents which during this season set southward to the Maldives. In November, December, and January, having passed these islands, they should steer so as to pass near Socotra. In February they should steer West, keeping on the parallels of 11° or 11° 30′ N. lat. In the end of March or beginning of April it will be better to keep on the parallel of 9° or 10°. In April northerly and N.W. winds are generally found, when they should keep their wind, making perhaps a few short tacks, so that they should not get too far South. The boards should be short, as the object is to make westing. In April it is not necessary to go near the equator; while in May, at the beginning of the S.W. monsoon, on the contrary, a vessel should get to the southward.

At the end of April and beginning of May, when the ship is only 2° or 3° West of the African coast, S.W. winds will probably be found, and veering to southward near the land. At this period a vessel should endeavour to make the land to the southward of Cape Guardafui, for if she falls to leeward of Socotra the passage will become uncertain and it may be necessary to cross the line in order to get to the westward.

From Ports on the East Coast of India to the Red Sea.

On leaving any port on the eastern coast of India for the Red Sea before the month of April, a vessel should pass along the southern coast of Ceylon; from thence she must take the Nine Degrees Channel and continue her voyage as previously shown for vessels leaving the ports of Cochin and Anjenga.

In April westerly winds prevail off the S.W. coast of Ceylon, and it would be difficult to coast this island and to reach the Nine Degrees Channel. After the month of March, therefore, a vessel in the southern part of the gulf of Bengal should take the South route for the Red Sea, keeping on the parallel of 9° or 10° S., and pass southward of the Chagos Archipelago, where in the beginning of the season more favourable winds for making westing will be found than by taking the direct route between the Maldives and the Speaker Bank.

The Little Strait of the Red Sea.—The Red Sea is entered by two channels, called the Great and Little Straits: the former between the cape of Babel-Mandeb and the island of Perim: the latter between this island and the Abyssinian coast. The Little Strait is much frequented on account of its depth of water admitting of anchorage when required. In approaching this strait the soundings suddenly decrease from 18 to 8 fathoms. A vessel entering with a fair wind, should keep mid-channel in the Little Strait, or rather nearer to Perim

Island than the Arabian shore. There is no danger in this channel, although the soundings are by no means regular, varying from 13 to 7 fathoms.

A vessel leaving the Little Strait and doubtful of reaching Mocha before night, with appearance of bad weather from S.W., should anchor under shelter to the northward of Cape Babel-Mandeb, in the entrance of the strait. There the sea is smooth; whereas outside, the sea will be found rough, and she will have greater difficulty from thence in reaching Mocha.

The Great Strait.—The Great Strait, which is about nine or ten miles broad, is bounded on the West by the coast of Abyssinia; on the South by the little islands called the Brothers; and on the East by Perim Island. It is bordered by a narrow ridge of soundings; and in the middle of it no bottom is found at 90 fathoms. At a short distance from the Brothers, and near the coast of Abyssinia, there is uneven ground, the soundings varying from 25 to 13 fathoms.

As no anchorage is to be had in the Great Strait, excepting off Perim Island, off the N.W. island of the Brothers, or near the Abyssinian coast, the Little Strait is generally preferred both on entering and leaving the Red Sea. With a good breeze, however, and during the night, the Great Strait is preferable, and it would not be prudent then to take the Little Strait.

A vessel entering the Great Strait should keep near Perim Island, where, in case of the wind falling and being set towards the Brothers, she may find good anchorage. On entering the Great Strait during the night, short boards should be made to the West of Mocha until daylight, when the coast of Abyssinia, in soundings varying from 15 to 19 fathoms, may be kept. It is better to make short tacks thus than to anchor with a fresh wind and risk the loss of an anchor. During the night a vessel should take care not to pass Mocha; for in the season of southerly winds the current sets strongly to the northward along the coast of Abyssinia. We may refer hereafter to the excellent directions of Horsburgh and Moresby on this subject.

From the Red Sea for India.—It is often difficult to clear the Red Sea, and vessels seldom attempt to do so from September to April, when easterly winds prevail in the gulf. In fact, a vessel then meets the N.E. monsoon, which is against her, either for India or the Persian Gulf.

In April, when westerly and S.W. winds are set in on the South coast of Arabia, a vessel may safely leave the Red Sea and proceed towards the Persian Gulf or the coasts of India. The favourable season for leaving it is from April to September, and a vessel then bound to Surat need not leave Mocha before the beginning of September, so that she may reach that place towards the 20th of the month, or at the end of the S.W. monsoon, before which it would be imprudent to anchor there.

On leaving the straits of Babel-Mandeb the course should be shaped East for the middle of the gulf of Aden, where the breeze is more constant than near the shore; and should the wind be light and variable the African coast should not be approached, by which calms and strong westerly currents will be avoided.

Bound to Ceylon or further eastward, the course

should be shaped for either the Eight or Nine Degrees Channel between the Laccadives and Maldives, a route which should be adopted from March to November, being then preferable to any other. It may even be taken during the height of the N.E. monsoon, provided Seuhelipar be passed close, if taking the Nine Degrees Channel. From October to March it is better to pass North of the Laccadives, and make southerly along the Malabar coast as far as Cape Comorin, and then run for Point de Galle.

For Muscat.—The most favourable season for the passage from the coast of Malabar to Muscat, or the entrance of the Persian Gulf, is from November to February. During these four months inclusive, a vessel leaving the gulf of Bengal, Ceylon, or the southern ports of the Malabar coast, should follow the coast, taking advantage of the land and sea breezes, which will be found as far as the high land of St. John, in lat. 20° N. She should then stand off the coast N.W. when the wind will permit, passing near the coasts of Guzerat, Cutch, Scinde, and Beloochistan. Having doubled Cape Diu, it will then be advisable to keep well to the northward, and, if the wind admits, to reach the latitude of 23° 50′ or 24° N., before gaining the meridian of 60° E. Indeed, even in this season, the winds frequently come from the northward. In crossing the gulf of Cutch, heavy squalls from East or E.S.E. are often experienced, with a dull cloudy sky, and, on the contrary, when the wind gets to the northward, the weather is fine and the sky clear; with N.E. winds white squalls are met, which are only indicated by a small cloud, and which would pass unnoticed but for the wind they bring.

When running West on the parallel of 24° N., a fresh breeze will most probably be found, until nearly half the Persian Gulf is opened, from the entrance of which violent squalls from N.W. must be expected. A vessel then proceeding up the gulf should keep the Persian shore: and if going to Muscat that port may be steered for, keeping a little to windward of it. At this season the passage is generally not more than ten or twelve days from Bombay to Muscat.

During this season a vessel making direct for the Persian Gulf from Bombay sometimes meets with northerly winds, which prevent her reaching the coast to windward of Ras-el-Gat, and if she nears the coast meets with calms. These can only be avoided by keeping five or six leagues from it.

In March, April, and May, the direct course for Muscat may be steered from any port on the Malabar coast, and as at this season the land breeze is not to be depended on, rendering it difficult to make northing, a vessel should stand out to sea, if she be to the northward of the Laccadives, taking the most suitable channel, if from any port on the southern coast of Malabar.

When well off the coast, the sea will be found smoother than near it, and the wind will be found varying from North to W.N.W., but most frequently to N.N.W. and N.W. With the wind from North, a ship would make her westing, and when from West, she would stand to the northward to reach Ras-el-Gat, if in the month of March or beginning of April, as speedily as possible.

In the latter part of April and in May a vessel should stand to the westward all she can, in order to profit by westerly and S.W. winds, which, as soon as the gulf of Aden is opened, will surely be met as the coast of Arabia is approached. She should then make for the coast to the South of Ras-el-Gat, off which southerly and S.W. winds are found during March and the beginning of April, blowing strongly also in May.

Within Ras-el-Gat, during this season, land and sea breezes are met, unless heavy squalls from N.W. come down from the Persian Gulf, which they do once or twice a month. During these months a vessel may keep at a distance of five or six miles from the coast, and passages from Bombay to Muscat are made in twenty days.

In September and October, at the change of monsoons, the passage is longer; the wind is then variable between the coasts of Arabia and Malabar. However, it is generally N.W. about Bombay and to forty or sixty leagues from the coast, off which a heavy sea is often found. At the end of September and during October a vessel should keep along the coast as far as the latitude of 19° or 20°, and then stand out to sea and profit by slants of wind, keeping to the northward of 19° N. in crossing for the Persian Gulf.

If bound to Muscat, she should endeavour to sight the land about Ras-el-Gat, whence she may have variable winds along the Arabian coast. If she is for the Persian Gulf, she should keep as far North as she can near the coast of Beloochistan, as far as Cape Jask, and avoid the Arabian coast.

Southern Routes from Bombay to Muscat.—The Southern route for Muscat and the Persian Gulf from Bombay, is a distance of 1,500 leagues, although the actual distance from Bombay to Ras-el-Gat is not more than 260 leagues. In June, July, and August, this southern route is generally followed by vessels to the Persian Gulf or the Red Sea.

On leaving Bombay they keep in soundings from 18 to 23 fathoms, working to the southward along the coast between the depths of 32 and 54 fathoms, not approaching the shore into less than 18 fathoms, and not standing seaward so as to lose soundings, especially in passing the Laccadives. Along the coast at this season strong winds from S.W. and W.S.W. prevail, sometimes accompanied by heavy squalls from West and W.N.W., with rain.

Having reached the parallel of 4° S. lat., and nearly on the meridian of the South point of Ceylon, in June and July, they fall in with the S.E. Trades; but in August these winds must be expected further South.

In the first two of these months, with the S.E. wind vessels may run West between the parallels of 4° and 4½° S., in order to pass between the southern extremity of the Maldives and the Speaker Bank. This is the first of the southern routes. In July it is better to pass South of Diego Garcia, and to keep in 9° or 10° S. lat. This latter route is generally safer than the former, because the wind is stronger and steadier than near the equator. However, vessels which have run West on the parallel of 4° S. in June, July, and the beginning of August, have found the sea smooth and the wind steady from E.S.E. to S.E. But from March to the end of

August, on the contrary, the first route is not followed because the wind frequently varies to West and occasions an easterly current. When the Trade wind prevails, a westerly current is generally found nearly all the year in the second southern route, and this should be preferred to that made on the parallel of 4° S., by vessels leaving the East coast of India for Bombay, the Persian Gulf, or Red Sea, as well as by those leaving Bombay for those seas.

Before leaving the southern latitude a vessel should be careful to make her westing, and therefore, if bound to the Red Sea, it will be best to pass near the Seychelles. When bound to Muscat or the Persian Gulf, she should reach to a degree or two West of Ras-el-Gat before leaving the Trades; for, during the S.W. monsoon, between the equator and the coast of Arabia W.S.W. or W.N.W. winds are generally found, which are always accompanied by an easterly current. Consequently, it then becomes impossible to make westing when once a vessel is North of the equator, so all she can do then is to endeavour not to fall off to leeward of North, for the cross sea met there often makes her do so. It is necessary then at least to gain the meridian of Ras-el-Gat, so as to reach it with a westerly wind. She would then sight this cape or the land South of it, but should not approach the dangerous gulf of Massera; but, having passed the island of this name, she would keep to windward as much as possible, in order to make the land.

In approaching Ras-el-Gat the S.W. winds, which blow strongly to the southward of the cape during this monsoon, gradually become S.E. as the cape is passed. When it remains at South the S.W. monsoon ceases and the wind is variable as far as Muscat. Once or twice a month strong S.E. winds are found near the cape, lasting two or three days and reaching to the middle of the gulf; nevertheless, N.W. winds prevail in these parts.

Direct Route from Bombay to Muscat.—The direct route from Bombay to Muscat is rarely taken by merchant vessels during the height of the S.W. monsoon. It has been adopted by ships of war in preference to the long circuit made in taking the southern route. In the direct route a vessel may work between the parallels of 15° and 19° N. lat. in order to make westing. Some vessels have made this passage without going further South than 13° N. lat.

Routes from the Persian Gulf to Malabar Coast.—Vessels leaving the Persian Gulf from September to April, during the season of the northerly winds, for any port on the Malabar coast, having doubled Cape Jask, should follow the coast of Beloochistan, keeping, however, at such a distance as not have to encounter the variable winds and alternate calms and breezes found near the coast.

In approaching the meridian of Cape Mouarree it will be found advisable to stand to the S.E., to cross the entrance of the gulf of Cutch, and then to pass at a good distance from Guzerat. When Cape Diu is doubled, at the distance of twelve or fourteen leagues, a direct course for Bombay may be shaped, if bound to that port, or for the high land of St. John, if bound to Surat.

A vessel bound to any port on the South coast of Malabar, or in the gulf of Bengal, should proceed as already stated, passing at the same distance from the coast of Beloochistan and that of Guzerat; but with N.E. and N.N.E. winds a vessel should keep far enough North to sight the peak of Barcelore, and then pass between the Laccadives and the coast.

In the S.W. monsoon, from March to September, a vessel leaving the Persian Gulf for ports on the western coasts of India, should keep in the latitude of Kundary, at thirty or forty leagues from the land, and then steer East for her destination.

Bombay is the only port on this coast that can be frequented during this season. The harbour of Surat is very bad, and vessels bound there will do well not to leave the Persian Gulf before the beginning of September. Then, preserving a good distance from the coast of Beloochistan, they should navigate so as to pass twelve or thirteen leagues from Cape Diu. When off this cape they will have attained the parallel of 20° N. lat., and may then steer eastward for the high land of St. John; they should look well out for the anchorage, and not get hold of the land too far North. By night, they should not approach it in soundings less than 13 fathoms, and when the land is sighted they should keep along the coast as far as Surat, in 9 or 10 fathoms water, until abreast of Damaon; they may then continue along the coast, in 6 fathoms at high water and 5 if low water, when going into Surat River.

In this season, if bound to the gulf of Bengal, a course should be steered so as to pass West of the Laccadives, and then between these islands and the Maldives, or through the channels among the islands, and then shape a course for Cape Comorin. At this season it is better to pass West than East of the Laccadives, as fresh winds and clear weather are found there, while to the East of them a vessel is exposed to squalls and rain in the whole of the S.W. monsoon.

CHAPTER XI.

NAVIGATION OF THE GULF OF BENGAL.

THE gulf of Bengal is of so much importance as to require our best attention. We propose therefore to consider the routes from the southern ports for those to the northward, and also those from the northward for the southern ports, during the two monsoons; and lastly, the routes from the gulf for the ports of the western coast of India, the Persian Gulf, Red Sea, &c.

S. W. Monsoon.—The S.W. monsoon is the most favourable season for leaving the southern and western ports of the gulf of Bengal for the northward.

From the Malabar Coast and Ceylon to Bengal, February and March.—On leaving the Malabar coast or the South coast of Ceylon for Bengal, in the end of February or beginning of March, with the wind moderate and currents favourable, a vessel would stand along the East coast of Ceylon as far as Baticola, and then steer for the coast of Orixa, so as to make it near the parallel of 19° N. lat. From thence she would follow the land as far as Point Palmyras, and afterwards take the most convenient route for the coast of Bengal.

In this route should a vessel encounter strong N.E. winds with contrary currents in the latitudes of the

Basses, on the S.E. part of Ceylon, she would make to the eastward as well as she could, giving them a good berth, for it is probable that at a certain distance from the coast she would find the wind between N.W. and S.W., and sometimes even S.E.

From the Coromandel Coast to Bengal, February and March.—During the same months, a vessel leaving the Coromandel coast for Bengal, will do right to give the shore a good berth, in order to profit by those variable winds which sometimes come from the southward; while in the vicinity of the coast easterly winds varying to N.E. are met with, which render it very difficult to make northing. When out at sea, if the wind permits, a vessel should steer so as to make the coast of Orixa about the parallel of 19° N.; otherwise as soon as possible she should sight the Jaguernaut or the Black Pagoda.

From Malacca Strait to Bengal, February and March.—During these months, a vessel from Malacca Strait should enter the gulf of Bengal by passing either northward or southward of the Nicobar Islands, according to the wind, and she would then steer for the coast of Orixa, as already observed.

Ceylon and the Coast of Coromandel to Bengal, from April to the middle of September.—When the S.W. monsoon is well established, that is, from the beginning of April to the middle of September, it generally extends to the West coast of the gulf. At this season, a vessel leaving Ceylon or the coast of Coromandel for Bengal, should not stand too far out from the land; for the wind sometimes veers westerly out at sea. Nor should she on the other hand keep too near the coast until she has passed northward of Vizagapatam, in order to avoid the strong sea breezes from S.E., which are found at the entrance of the large bays between Madras and that port. As the current sets strongly to the N.E. at this season, it will be advisable for a vessel, when uncertain of her distance from the land, to sight it in lat. 18½° or 19°. From thence she may steer for Bengal, following the particular directions (given by Horsburgh) about making the land of the Hooghly during this monsoon.

From Malacca Strait to Bengal, from April to the middle of September.—In the same month, a vessel leaving the strait of Malacca for Bengal, should take the outer passage, crossing the channel of Surat or that of Pulo Brasse if leaving Acheen, and that between Pulo Way and Pulo Malora, if from the strait of Malacca. This latter, according to Captain Millar, is the best she can follow from the strait of Malacca to Acheen, and for reaching the two other channels abovementioned. She will then pass West of the Andaman and Nicobar Islands. She may then take the channel between Pulo Way and Pulo Rondo, when from the strait of Malacca; or that formed by Pulo Way and Pulo Brasse, when from Acheen. In this case she will pass the Nicobar Islands to the eastward, and will then enter the gulf between these islands and the Little Andaman by the Ten Degrees Channel. She might also keep to the eastward of the Andaman Islands, and take the passage between the Great Andaman and the Cocos Isles, from whence she would make the coast of Orixa near Point

Palmyras, unless the wind should veer to W.S.W., and the current set N.E., which is often the case during the neight of the monsoon. For this reason, a vessel leaving Acheen or the strait of Malacca, should take the passage West of the Andaman and Nicobar Islands.

From Ceylon or the Coromandel Coast to Bengal, end of S.W. and beginning of N.E. Monsoons.—The route to be taken from Ceylon and the coast of Coromandel to Bengal, after the 15th of September, and during the month of October, is very different from those previously mentioned, on account of the termination of the S.W. monsoon, and the winds veering frequently to N.E. A vessel should then keep well off the coast as far nearly as the middle of the gulf. If she finds S.W. winds still there, she should steer direct for Point Palmyras: if, on the contrary, the wind is variable and from northward or N.E., she should take advantage of every favourable slant to make to the eastward, and when about twenty or thirty leagues from the Andaman Islands or Cape Negrais, she should tack to the N.W., and make all the northing she can. This will be by keeping out at sea rather than near the coast of Arracan, unless it be in October and November, when she may keep this coast.

A ship from Pondicherry or Madras at this season, with light variable easterly winds, may be drifted by southerly currents to the East coast of Ceylon before she is fairly out to sea, that she must be careful to guard against. N.W. winds, favourable for going to the East coast of the gulf, will be frequently found at sea. In the beginning of the N.E. monsoon it will be necessary to keep off the coast of Orixa, to avoid the southerly current, which is then very strong. If at this period the land be made by a vessel to the southward of Point Pamyras, she would risk not being able to make the voyage, or at least she would be much delayed, on account of the difficulty she would have in getting to the northward against the southerly current, which attains its greatest strength in November and December. It is therefore more advantageous, even at the end of December, when the currents begin to be less felt, to keep out to sea in the bay of Bengal, or nearer the eastern coast than the West, in order to make with certainty for the bank of Saugor or the Eastern Sea Reef. On the West coast, at this period, the currents sometimes run at the rate of three miles an hour. When it is desired to make the land North of Point Palmyras, a vessel should avail herself of the tide to double that point. She may anchor at pleasure, and need not continue under way when her boards are no longer favourable.

During the N.E. monsoon, on leaving the coast of Coromandel for Bengal, a vessel should make for the East coast, and make her northing along it or in the middle of the gulf. This method, indeed, is generally preferred, as the southerly currents, which often prevail at this period on the East coast, are thereby avoided. A similar proceeding may be adopted from the bay of Bengal, in order to gain the channels of the Hooghly.

From Acheen or the Strait of Malacca to Bengal.— During the N.E. monsoon, a vessel leaving Acheen or the strait of Malacca for Bengal, should adopt that route which the wind or existing circumstances admit of.

She may pass either East or West of the Nicobars, when entering the gulf of Bengal, by one of the channels between these islands and the little Andaman. Nevertheless, in leaving the strait of Malacca Horsburgh thinks it is preferable to make for the island of Narcondam after having taken a departure from the South point of Junkseylon; then passing North of the Great Andaman, between it and the Cocos Islands, or through the channel formed by these latter and Preparis. The starboard tack would then be adopted for entering the Hooghly, and some boards made should the wind be variable. A vessel should always be careful to keep at a good distance from the coast of Arracan, and, indeed, from the whole of the eastern coast of the bay of Bengal.

During the N.E. monsoon the wind in the gulf of Bengal is generally moderate and frequently changes to East. During this monsoon the sea is generally smooth. A vessel entering the gulf about the 1st of February, should lose no time in making to the eastward, for the East and S.E. winds begin about this time on the coast of Ceylon and North of the Basses. Besides this, the current sets along the coast of Coromandel, and it is therefore advantageous to keep near the land. In the strength of the monsoon, on the contrary, it will be better to keep out at sea, and work in the middle of the gulf, or on its eastern coast. Towards the middle of September, when the pilots of Calcutta move their station from Point Palmyras to the Sand Heads, a vessel should make to the eastward in the bay of Bengal and gain soundings to windward of the light-vessel. The current runs with incredible rapidity to the westward towards the end of September and in October; and if a vessel gets to leeward, she will have great difficulty in making the Saugor Channel.

From Bengal to the Coromandel Coast.—The most favourable season for making a quick passage from Bengal to the coast of Coromandel, or ports in the southern part of the gulf, is from the middle of October to the middle of February, when the N.E. monsoon is at its height.

A vessel leaving Bengal in September for Pondicherry, Madras, or any other port on that coast, meeting light winds from S.S.W. to West, and a slight northerly set, should keep at a little distance from the West coast of the gulf. In case of meeting southerly winds she should keep along the West coast, and afterwards work to S.W. without leaving soundings if she can. But if she cannot make progress by this method she should stand well out to sea, and avail herself of every slant of wind. The currents, however, at this season are very irregular, for while they are setting strongly to the N.E. near the coast, they have very little strength out at sea; and in this month, when a N.E. current prevails at sea, it is not always to be found near the coast. During this month, according to circumstances, a vessel may make the shortest passage as often by keeping near the West coast as out at sea. A vessel, however, should avoid getting so far to the eastward as the middle of the gulf in making her passage to a port on the coast of Coromandel during September, on account of the westerly winds which prevail there at this season.

A vessel from Bengal to any port on the Coromandel coast, between October and December should keep off the coast, and avoid the gusts of wind met with at this season. After the middle of October N.E. winds may be found on leaving Bengal, often lasting several days; but the further South the more variable is the wind, which at a distance off the coast veers to S.W., and near it becomes East.

Whatever may be the port on the Coromandel coast to which a ship is bound after the 15th of October, the land must be made about forty miles North of it. The same applies to any port on the East coast of Ceylon, on account of the strong southerly currents; and Palk Bay should be avoided on account of the westerly current at its entrance. The coast is consequently dangerous with a N.E. wind.

A ship leaving Bengal in January for ports on the Coromandel coast, should keep at a mean distance from the land, beyond the region of the light variable winds, and should steer direct for her port when a little northward of its latitude. But with the wind fresh at N.E., it will be better to shape a course North of it.

Adopting the same route in February, she should give the Orixa coast a good berth as soon as she clears the Hooghly, and should steer South in order to avoid the calms and light breezes of the land; while out at sea, and in the eastern part of the gulf, the wind will be found fresh from N.E.

In the beginning of February, on the coast of Coromandel, S.W. winds are generally found, with currents setting northerly. The coast must then be avoided, and only made to the southward of the port bound for. However, if the wind continues from North or N.E., a vessel might steer for her port as soon as she is in the latitude of it, taking care from the 1st of February not to be too far North.

Ceylon.—During the N.E. monsoon, from September to March, a ship bound to Ceylon should sight the land North of her port of destination if it be on the East coast of this island. If bound for the western or Malabar coast, she should sight the island in the latitude of the Aganis or South of the Basses, and then steer along the South and S.W. coast of it.

From Bengal to the Strait of Malacca.—Leaving Bengal for Acheen or the strait of Malacca during the N.E. monsoon, a ship should steer S.E. from September to May, passing between Cape Negrais and the island of Preparis, or else between this island and the Cocos. From thence she should make for Pulo Way if bound for Acheen, and for Pulo Bouton or Prince of Wales Island if bound to the strait of Malacca. In all cases she should be careful of currents setting South and S.W., which are found on the eastern side of the bay of Bengal, when taking the channels between the northern islands of the Andaman Group. A vessel West of these islands should steer West of the Great Andaman, and, if the current permits, should endeavour to enter the Duncan Channel, between the Great and Little Andamans. In the opposite case, being to the eastward, a vessel should endeavour to reach the Ten Degrees Channel, between the latter island and that of Carnicobar. The passage will then be much longer, and it will be advisable as soon as clear of the Hooghly to steer as much to the eastward as possible, so as to pass North of the island of Preparis. Between this island and Junkseylon during the N.E. monsoon the current is very variable. In the beginning of the monsoon it generally sets N.W.; in March and April, South or S.W.

From Bengal to the Coromandel Coast.—A ship leaving Bengal in March and April for ports on the coasts of Coromandel and Ceylon will have winds varying from from S.S.W. to West; against which she must make all the southing she can, steering South or S.S.E. when she can do so. In the middle of the gulf at this period light winds, varying from North to West, sometimes prevail, while near the Coromandel coast the wind is S.W. To profit by the northerly breezes found during these two months a vessel should keep East of the meridian of Point Palmyras until some southing is made, and should be careful, especially towards the end of April, not to approach too near the large Andaman Island, in the event of a westerly gale coming on and making it a lee shore. But a ship should take every advantage of the changes of wind to make southing, and in crossing the gulf should get at least sixty miles to the southward of her port of destination before steering for it, and be prepared for the northerly current with the southerly winds found near the coast.

Eastern Route from the Andaman Islands.—In the same season of the year, between the Andaman Islands and the East coast of the gulf, N.W. winds are generally found; and it is on this account that many commanders of ships on leaving the coast of Bengal for Europe or ports on the West coast of India steer East of these islands, passing northward or southward of the Nicobar Islands as most convenient. Horsburgh thinks that whenever the wind admits it will be better to pass West of the Andamans, and that towards the end of April it will be most advantageous to work to the westward of this group in order to proceed South.

Ceylon for the Coromandel Coast.—A ship bound to Ceylon at this period should, as before mentioned, keep out to sea till she has reached the parallel of 10° N., on the meridian of Point Palmyras. She may then make for the coast, taking care to keep South of her port. If bound to the West coast of Ceylon or Malabar, when she has reached the latitude of 9° she should steer so as to near the South part of Ceylon, and should not sight the land North of the Basses. During these two months, when bound to the Malabar coast, she should keep off the West coast of Ceylon, and, indeed, well out at sea, so as to double Cape Comorin without difficulty.

Bengal to the Strait of Malacca.—A ship from Bengal bound to Acheen or the strait of Malacca during the S.W. monsoon should take nearly the same route as during the N.E. monsoon. She should steer S.S.E., if the wind permits, until she is South of lat. 15°, steering from thence towards the Cocos Islands when certain of her position, or, if preferred, for a landfall somewhere near the North point of the Great Andaman. A vessel not certain of her position will do better to gain the parallel of 14° S. before bearing up for the Cocos Islands, in order to pass to the southward of these islands. After passing them she should keep her wind in making to the southward, in order not to approach the islands off the coast of Tenasserim, and she should also keep clear of the Andaman Group. She should then, if possible, steer directly for Barren Island, visible at the distance of thirteen leagues; and, having passed it on that side which seems most advantageous, she should keep to windward if bound to Acheen, but if for the straits it will not be necessary to keep so close a luff. Nevertheless, southing must be made in order to pass at a good distance from the Sayer Islands, at the South extremity of Junkseylon, in case of S.W. winds prevailing in these parts, one that does not often happen during this monsoon. From the South point of Junkseylon she should steer direct for Prince of Wales Island.

In order not to be thrown to the eastward, on leaving the Hooghly between April and September a ship should run along the West coast as soon as possible, if the wind permits, as far as the neighbourhood of Point Palmyras. Probably during the night the wind will veer a little towards land, and will be favourable for steering South, while during the day it veers seaward. In order, therefore, to profit by these changes, a vessel should be near land between two and three o'clock in the morning. In June and July these breezes are often pretty strong; but they are not always found, and if the ship has the regular monsoon it is still preferable to be near the coast instead of in the middle of the bay, where the sea is always highest. On this account, therefore, she should not get further off shore than will enable her to make a little southing.

In the case of a vessel near the coast of Orixa meeting with contrary winds, which will not admit of her getting to the southward, she should make to S.E., so as to get within sixty or seventy miles of the land, where she would find less current. It would not be prudent to stand farther out in the hope of meeting with a better wind. By keeping along the land or at a moderate distance to sea from it, and taking advantage of the changes of wind, for a ship is more likely to find the land breeze as she advances to the southward, she will find less difficulty in working to windward than in keeping out at sea.

On reaching Point Gordeware the coast should not be approached South of this point on account of the bays between it and Pulicat, but work to the southward, keeping at a convenient distance from the land. From Pulicat southing may be made with the land breezes which come off during the night and the S.E. winds which prevail during the day, by keeping near the Coromandel coast.

If bound for Trincomalee the ship should continue to work near the West coast of the gulf as far as Negapatam, and then stand out to cross the opening of Palk Bay, so as to reach the coast of Ceylon.

Route down the Gulf of Bengal.—A ship coming down the gulf of Bengal in the S.W. monsoon may follow the above directions until the middle of August, and the route is generally along the West coast of the gulf. Nevertheless, when she is bound for the West coast of India or for Europe, it is not necessary to keep so close to it as above. If, on leaving Bengal, she has a good westerly wind, she should take advantage of it, steering South; and should the wind change to S.W. or S.S.W., she should make some boards along the West coast, and in standing to the eastward should be careful of the Andaman Islands. If she can work easily along the West coast, it will be better to keep to this route; if, on the contrary, there is difficulty in doing so, she should tack to S.E., and take advantage of every change of wind to gain southing, keeping then in the middle of the gulf. When the wind admits of passing the Andamans about thirty or forty leagues West of the Little Andaman, she should continue the S.E. tack, for it is probable that she may also be able to clear the Nicobar Islands and Cape Acheen on the other tack to the westward.

Captain Miller does not quite agree with Horsburgh; indeed, he says that on leaving Bengal a ship should take the S.E. tack, so as to pass West of the Andaman and Nicobar Islands. In case this cannot be done, he says, that instead of losing time in tacking westward, it will be more advantageous to pass North of the Andaman Islands, through one of the channels of the group—the principal one of the islands being the best. Then, having gained to the East of the Andamans, she should still stand southward until she loses the S.W. monsoon under the lee of Sumatra. Near the coast of this island the wind will be found variable, often favourable for again taking the gulf by the strait of Surat or that of Pulo Bras, passing between Pulo Way and the island of Malora.

Strait of Malacca.—A ship leaving Acheen or the strait of Malacca in the S.W. monsoon should pass through the strait of Surat if the wind admits. It will be better, says Horsburgh, to double the North extremity of Pulo Bras close in shore, where westerly currents are often found, while at the same time they set N.E. between Pulo Rondo and the Nicobar Islands. Consequently, a ship should not make for the gulf of Bengal by the great channel. A vessel leaving the gulf of Bengal for the westward should not go too far East nor enter the zone to the southward between the parallels of 3° and 4° N., for here the wind is very light, but attended with frequent squalls from S.W., followed by calms. As soon as she has gained a little West of Cape Acheen, she should endeavour as soon as possible to reach the S.E. trade, giving a good berth to the islands off the West coast of Sumatra; and this she will soon accomplish if she takes the most favourable tacks according to the changes of the wind.

With regard to the general rules for sailing in the gulf of Bengal, according to the season, Captain Miller gives us the following in the *Nautical Magazine* for 1843.

From 15th Jan. to 31st May.	Going North or up the gulf, take the western side; coming South or out of it take the eastern side.
In June, July, and August.	Going North or up the gulf, keep in the middle of it; coming South or out of it, take the eastern side, even East of the Andamans.

In September, October, and November.	Going North or up the gulf, take the eastern side; coming South or out, take the western side.
In December and to the middle of January.	Going either North up the gulf, or coming South out of it, keep the middle of it and make small tacks.

From Ports on the East Coast of India to those on the West.—A vessel from Tenaou, Madras, Pondicherry, or Karikal for ports on the West coast of India, the Red Sea, or Persian Gulf, should adopt the following method of leaving the gulf of Bengal. In leaving Tanaou, if the land and sea breezes are well established, she should take advantage of them as far as Pondicherry; but if she cannot do this, she should keep out at sea, as in leaving Pondicherry. She should then make good a S.S.E. course or thereabouts, with the W.S.W. or West winds which are often found at sea. Near the equator the S.W. monsoon decreases and the winds generally veer to the South; she may then make to the S.E. on whichever tack she can best do so. According to the season, two routes are to be taken for ports on the West coast of India—the Northern and Southern Routes.

Northern Route.—The northern route is the most direct, but it should only be taken towards the end of June, July, and the beginning of August, when S.E. winds prevail near the equator. After crossing it to the West, keeping as much as possible in the prevailing winds, as soon as she finds the S.E. wind steady, which at this period is often the case in 4° or 5° S. lat. she should steer West between these parallels or on the latter; but on approaching the meridian of 69°, she should not keep to the southward of 4½°, in order to pass the North part of the Speaker Bank of the Chagos Archipelago. Having passed it, she should keep her westerly course, and cross the equator on the meridian of 58° or 59° E.

Southern Route.—The southern route is more certain than the northern one during the whole of the S.W. monsoon. However, as in taking this southern route observations cannot sometimes be had, owing to cloudy skies and squally weather, many seamen prefer the northern route. Vessels obliged to cross the equator well to the eastward, on leaving the gulf of Bengal, Sumatra, or Java, should always take the southern route.

In taking this route a ship should endeavour to reach the S.E. Trade as quickly as possible, and that part where it is well established at this period, between the parallels of 9° or 10°: from whence she should steer West, and if not sure of her position she should sight the island of Diego Garcia, and then steer S.W., so as to pass South of the Centurion Shoal. In case of not sighting that island, she should steer West between the parallels of 8° and 9° taking care to avoid the shoals South of the Chagos Archipelago. She should then return northerly again, and cross the equator on the meridian of 58° or 59°. As in the northern route, whatever her destination may be, before returning into northern latitude she should stand sufficiently to the westward to reach it with the Trade wind, which is sometimes very strong from West, producing a strong easterly current.

Persian Gulf.—If bound to the Persian Gulf she should cross the line on the meridian of 49° or 50° E., and then follow the directions given in a preceding page (146).

Red Sea.—A vessel bound to the Red Sea, after having reached 65° E. long., should steer a little to the northward of West. She should pass near the northernmost of the Seychelle Group, and if the weather permit should sight Dennis or Bird Island. From thence she would cross the line between 43° and 45°; then make for Cape Guardafui and sight the land some distance South of this cape, for if she be thrown to leeward no ordinary sailing vessel can ever recover it. She should then make way along the African coast as far as Mait or Burnt Island. She may even keep along the coast further West, on account of the prevailing westerly winds, before standing over for Cape Aden, and even to Ras Amran if the winds allow her.

If going to Bombay, Mahé, or any port on the West coast of India, Horsburgh says cross the line between the meridians of 60° and 61°, for during the height of the monsoon strong westerly and W.S.W. winds prevail in the Arabian Gulf, with a heavy sea. As a general rule, therefore, a vessel should not cross the line so far East during the height of the S.W. monsoon as to oblige her to cross it obliquely in order to reach her port of destination.

In running to the northward for Bombay a vessel

should give the Laccadives a wide berth; nor should she approach the land before reaching the latitude ot the island of Kundary. If going to Mahé, Calicut, or Cochin, she should take the Nine Degrees Channel.

Remarks.—It may be observed here, that from April to October the West coast of Hindostan is rendered dangerous by the bad weather. The passage from the coast of Coromandel to the Malabar coast, is generally made between October and January; and from the latter for the gulf of Bengal, the strait of Malacca, Acheen, Bantam, Batavia, &c., from February to April.

Leaving the Malabar coast for Europe, some time in the course of December or to the middle of January is adopted, especially when it is desired to touch at the Mauritius or Reunion. A vessel sailing later than this would find great difficulty in passing the Cape of Good Hope. It may be as well to state these times for sailing, as it is useful to know them, in order to make the proper passage. And we will conclude these remarks on the gulf of Bengal with some passages from ports on the West coast to those on the East and *vice versâ.*

From Ports on the East Coast of the Gulf to Rangoon.—From Bengal, during the N.E. monsoon, a vessel for Rangoon should sight Cape Negrais, and keep South of the Alguada Reef, which extends along the cape in this direction. During the same season, from Pondicherry or ports on the Coromandel coast, a vessel should make to the northward and then take the channel North of the island of Preparis, or else the South channel of this name, according as she finds it most convenient. At this season it will be better to make easting when the reefs of Cape Negrais are passed, and to keep near the coast with the lead going, for there is very little tide, and the current from the rivers generally sets S.W. or N.W.

Vessels from the strait of Malacca, Acheen, or the Nicobar Islands, bound for Rangoon during the N.E. monsoon, should sight the western of the Mergui Islands, and then make to the northward for the entrance of Rangoon.

The routes just mentioned also avail for ports on the coast of Martaban.

Leaving Bengal for Rangoon during the S.W. monsoon, if the wind permits a vessel should endeavour to make the island of Preparis or, what would be still better, Cocos Island. She would then take the most convenient channel, navigating so as to make the coast of Pegu, a little West of the bar of Rangoon. In the same season, when leaving any port on the Coromandel coast, a vessel should endeavour to make Landfall Island, (near the northern point of the large Andaman,) if the wind hangs to the southward, or Cocos Island if the wind draws westerly, and she should then pass through the channel between them. On leaving Cocos Channel she should steer East to sight the island of Narcondam; and then steer N.E. to make the coast of Pegu, a little West of the bar of Rangoon. If the land is made to the eastward she would make up her westing with the tide.

A vessel from Acheen or the strait of Malacca in the S.W. monsoon, should sight the island of Narcondam, and from thence steer for Rangoon bar.

From Rangoon to Ports in the Gulf of Bengal.— From Rangoon during the N.E. monsoon for Bengal, a vessel on clearing the river should shape her course so as to pass outside all the shoals near the coast of Pegu, thus passing southward of the dangers off Cape Negrais. She would then continue along the coasts of Ava and Arracan, in order to reach the Hooghly.

If bound for some port on the Coromandel coast, she should take one of the channels between Cape Negrais and Landfall Island, and then proceed for her port, making the land to the northward of it before February, and to the southward after this month.

In the same season, a vessel going to the strait of Malacca, should sight the southern extremity of Junk-seylon. If for the Nicobar Islands or to Acheen, she should steer direct for the port of destination.

On leaving the bar of Rangoon for ports in the gulf of Bengal during the S.W. monsoon, a vessel should keep her wind in standing out to sea, whether for Bengal or a port on the Coromandel coast, Acheen, or the strait of Malacca. She should not discharge her pilot until she is well out to sea, with Elephant Point to the northward. When it bears N.b.W. from 4½ fathoms, she will be about mid-channel. From thence standing out to sea, the soundings will decrease to 4 fathoms, and she should tack and stand in for the land to 5½ fathoms. She would then tack again, and by increasing her depth ascertain that she is approaching the shoals which lie off the coast.

When she has brought the wood of China Bucker to bear W.b.S., she would tack out to sea. It will, however, be advantageous to keep near the coast, in order to anchor if requisite, and to profit by the tides, which are very strong. When she has a depth of 5 fathoms, the wood of China Bucker should bear W.b.N. distant twelve or fifteen miles. If the night comes on before getting off China Bucker, she may drift with the tide to windward, keeping her lead constantly going. After the River Dalla is passed, she should not approach the coast nearer than in 6½ or 7 fathoms, as far as Baragu Point, which is indicated by the bottom of sand and shells. A good sailing vessel may continue her course westward, passing between the Cocos Islands and Preparis, or through the channel North of Preparis, from whence, if she is bound to Bengal, she should make for the harbour of Balassore, provided the wind only varies from S.W. to S.S.W., otherwise she should tack, in order to profit by the change of wind that she may make her westing.

If bound to Madras during the S.W. monsoon, she will find great difficulty in crossing the gulf after leaving the coast of Pegu, and vessels which are not good sailers should perhaps not attempt it. It will be better, with such vessels, to follow the same route as that taken from Rangoon to Acheen, thus, after profiting by the tides, to proceed West, from Point Baragu, when bound for Madras or any port on the Coromandel coast from Acheen or the strait of Malacca. The vessel should keep out to sea if the wind becomes westerly, and should endeavour to sight the island of Narcondam. If she is going South, some boards should be made, in order to keep the islands situated near the coast of Tenasserim at a respectable distance, although there are several safe channels through them. If she is going to the strait of Malacca, after she has passed the South point of Junkseylon, she should make direct for Prince of Wales Island. If bound for Acheen, she should if possible get to the westward, towards the islands of Nicobar, or else keep the southern tack till she reaches the coast of Pedir, (North coast of Sumatra,) where she will find the current running West, which will be favourable for reaching the harbour of Acheen.

A course may then be taken for the gulf of Bengal between Pulo Way and Pulo Brasse by a vessel bound to any port on the Coromandel coast, taking care while out at sea to profit by every change of wind to gain westing. Weatherly vessels have made this passage in fifteen days in the strength of the S.W. monsoon.

These routes have been here described in detail because they afford a general idea of the navigation which should be adopted in all seasons by vessels passing to or from ports on the coasts on either side of the gulf of Bengal to the opposite. The following will complete our details.

Coromandel Coast to Mergui.—Vessels from the Coromandel coast bound to Mergui, between the 1st of March and the middle of September, are obliged by the westerly and S.S.W. winds to pass to the southward of the Little Andaman, after having reached the parallel of 10° 10′ or 10° 15′. If the Ten Degrees Channel were crossed within these parallels they would be able to sight this island and thence steer with more confidence for the eastern coast, which should be made about the

Torres Islands. Although it may be unusual, especially at this season, to meet with westerly currents, it will be advisable when the Little Andaman is not seen to make good some thirty leagues beyond the reckoning before standing to the northward, in order to avoid the error of falling in with this island during the night when it was supposed to have been passed. In the same manner a vessel on leaving the coast of Coromandel may steer for the Sombrero Channel. D'Après considers this special precaution useless, and he thinks a vessel would be far enough to windward by taking the Ten Degrees Channel. In the passage from the Andaman Islands to the eastern coast of the gulf, vessels are often set to the eastward by the current, and therefore it should be provided for.

The islands of the Mergui Archipelago are high, and soundings may be had in the offing that will prevent all danger. When these are made a ship should steer for Isle Cabosa; then, to enter the archipelago by steering to the eastward, she may pass either North or South of this island, leaving the western Canister Island to starboard, and then follow the directions if bound to the port of Mergui.

Mergui to the Coromandel Coast.—The departure from Mergui for the Coromandel coast should not be later than the 15th or 20th of December, in order to reach it in the beginning of January—the season a vessel may make it in perfect safety. The N.E. monsoon being at this season in full force, a vessel leaving Cabosa Island should pass between Preparis and Cocos or between the latter and the Great Andaman. From thence she should run for the Coromandel coast, taking care to make the land North of the port to which she is bound.

For a vessel not leaving Mergui before the beginning of February, the safest route is to pass South of the Little Andaman. At this season she should profit by the winds, which in the gulf of Bengal blow more frequently from the South than the northward.

A vessel leaving Mergui during the S.W. monsoon for ports on the gulf of Bengal, Acheen, or the strait of Malacca, should follow the same route as when going from Rangoon to these ports during the same monsoon.

Madras to Acheen.—The most favourable season for leaving the Coromandel coast from Madras for Acheen, is between the 15th of August and 15th of September, that is, towards the end of the S.W. monsoon. During this monsoon, and more particularly between the 15th of August and 15th of September, M. D'Après states that a vessel having reached the offing would meet W.S.W. winds, varying to S.S.W., by which she would profit to make the island of Sumatra on the parallel of 5° N. She will thus be to windward of the channel of Surat, which, although narrow, is the best in going to Acheen. Captain Carnegie says that on leaving Madras during the S.W. monsoon for the strait of Malacca, it is better to take the Ten Degrees Channel than keeping to windward for Acheen.

Madras to Acheen and Malacca Strait.—During the N.E. monsoon and at its commencement, the same captain tells us that a vessel on leaving Madras should sight Sumatra South of Acheen, and then take the Surat or Bengal Channel according to whether she is going to Acheen or the strait of Malacca. She would then work along the Sumatra coast as far as Diamond Point, when from this point she would lay up for Prince of Wales Island.

Towards the end of the N.E. monsoon or after the 15th of February, when southerly winds are becoming established near the land, it will be advantageous to make northing along the Coromandel coast and then stand out to sea, so as to pass between Cape Negrais and the Great Andaman, or else to take Duncan Channel, between this island and the Little Andaman.

In these passages northerly currents are seldom found, but more frequently those setting southward. When observations cannot be had it is safer to consider that the vessel is set to the eastward.

Malacca or Acheen to Bengal.—A vessel leaving the strait of Malacca or Acheen during the S.W. monsoon for Madras, should keep near the North coast of Sumatra, (the coast of Pedir,) where she will find a westerly current, with land breezes. She would also pass Acheen Roads, and if the weather is bad may anchor there if necessary. Thence she should make for the Bengal Channel; from whence she would easily weather the Nicobar Islands, and would have a fair wind for the ports on the West coast of the gulf and even Calcutta. If bound for Madras, and the wind draws westward, she would probably have to make some tacks in order to gain her destination; but in no case should she cross the line. When she is working to the westward it will be better to stand in for the Coromandel coast to the northward, and afterwards to work along the land for her port.

Malacca to Madras.—As the weather on the Coromandel coast during the N.E. monsoon is bad, it is better not to leave the strait of Malacca before the 10th of December. From Malacca a vessel should shape her course for Pulo Pera, taking care to avoid the dangers in the middle of the strait of Malacca. From Pulo Pera, in going to Madras, she should steer for the Nicobar Islands, passing either between these islands or South of them. Some captains prefer the Sombrero Channel and others the Ten Degrees Channel. If she going to Masulipatam or Tanaou, after leaving Pulo Pera she may steer a course to pass South of the Cocos Islands. She should then make for her destination, keeping well to the northward of it, especially in January, so as not to be set past it by the current.

Malacca to the Malabar Coast.—A vessel leaving the strait of Malacca towards the end of October and November for ports on the West coast of India, from Pulo Pera should shape her course South of the Nicobars, and should then stand to the westward, in order to make the coast of Ceylon in the latitude of Aganis. She should then stand along the South coast of Ceylon and haul up for Cape Comorin.

Ceylon or Coromandel Coast to the Strait of Sunda.—A vessel from Ceylon or any port on the Coromandel coast in the S.W. monsoon for the strait of Sunda, will find the S.E. monsoon prevailing South of the equator. She must, therefore, in making her easting, keep as much East as possible North of the line. She should steer so as to cross it in 93° or 94° E., and it is probable that with South and S.S.W. winds in squalls she would be far enough to the S.E., or nearly so, to pass at a proper distance from the islands on the West coast of Sumatra, tacking to the S.W. occasionally with the wind from S.E. In this route northerly currents are common; nevertheless a vessel taking advantage of the shifts of wind may easily get to S.E., seeing that near these islands the wind often becomes South and S.S.W., while out at sea the S.E. wind prevails. She should therefore cross the equator well to the eastward, as above said, and keep at a moderate distance from the Sumatra coast.

If she is bound to Bencoolen she should get on its parallel before steering for it, passing to the northward or southward of Trieste Island, according to the prevailing wind. If the vessel is going direct to the strait of Sunda, reaching the entrance of the strait, she should keep off the land, for it is easier then to work to windward than by keeping near the shore. Nevertheless, in all seasons a good sailing vessel may make her passage from Bencoolen to the strait and back by keeping in with the land.

In all seasons in coming from sea it is prudent to make direct for Cape Java if desiring to take the strait of Sunda, and a ship may borrow to the westward or eastward as she approaches the entrance of the strait, according as she has the wind or current.

A ship leaving Ceylon or any port on the Coromandel coast for the strait of Sunda while the N.E. monsoon prevails North of the equator will find the N.W. monsoon South of it, and must take nearly the same route as that pointed out for the other monsoon. She should get to the eastward in order to cross the equator on the same meridian.

Leaving Ceylon in the month of October, the season when the N.W. monsoon is not always found South of the equator, a vessel should stand out to sea if she has the wind from N.E., in order to make some degrees of easting before crossing the equator with it, which is met with from November to March on leaving Ceylon.

In December, January, and part of February we generally find the N.W. monsoon fresh South of the equator, and a vessel may steer direct for the strait of Sunda as soon as she clears the S.E. part of Ceylon.

Coromandel Coast for Eastern Straits.—In leaving the coast of India for the eastern straits during the S.W. monsoon a ship should follow the route already pointed out for the strait of Sunda, and she will find the S.E. monsoon South of the equator. She would then get on a bow-line for the strait of Sunda against the S.E. monsoon, which is very strong in June, July, and August, producing westerly currents. It will therefore be necessary to make long boards to the southward at this time, while about the beginning and end of the monsoon the wind is variable.

During the N.E. monsoon a ship leaving any port of India should take about the same route for the eastern straits as for the strait of Sunda. When bound to those straits East of Java in the height of the N.W. monsoon, which prevails South of the line, she should sight the island of Noussabarou, particularly if she be uncertain of her position, for in November and December cloudy weather is often found on the South coast of Java, along with strong winds and easterly currents. If the land is made to the West of Noussabarou she will, in all probability, find finer weather by keeping four or five leagues off shore than by keeping out to sea.

Sunda Strait for Madras.—A vessel leaving the strait of Sunda for Madras or ports on the Coromandel coast during the S.W. monsoon should steer for the mountain called the Friar's Hood, on the East coast of Ceylon; then, running to the northward, should take care not to make the coast North of her destined port.

In the same season, when she is bound to the western coast of India, she should run to the westward with the S.E. Trade, and take the southern route, between the Chagos and Seychelles.

A vessel finding herself near the equator in April or October, when the N.W. and northerly winds prevail in the northern hemisphere at the change of monsoon, if the wind inclines westerly may run for any of the ports on the coast of Malabar and reach Mahé. If she is bound to Bombay she should stand to the northward to the East of the Maldives. But should she be some degrees West of these islands before reaching the equator, it will be better in April, the end of September, and October to pass them on the West, in order to reach that port.

During the N.E. monsoon, in returning from the strait of Sunda to the eastern coast of India, a vessel should steer so as to pass Hog Island or Cape Acheen about sixty or eighty leagues distant, and then follow the directions given for Route No. 6.

In the same season if she is for the West coast of India she should cross the equator between the meridians of 82° and 81° E. In the vicinity of the equator, in fact, and in a space to the northward of it, westerly winds are frequently met, with which she may reach Point de Galle. Thence she may cross the gulf of Manar and proceed up the Malabar coast, remembering what has been said concerning the winds and currents on the western coast of India, and the navigation of that coast.

CHAPTER XII.

ROUTES TO CHINA BY THE CHINA SEA AND THE CHANNELS EAST OF BORNEO.

THE straits by which the Indian Ocean is left for the China Sea are those of Malacca, Sunda, and the eastern straits of Bally, Lombock, Allass, Sapy, Flores, Aloo, Panthar, and Ombay. The several routes from the Cape of Good Hope to these straits have been already alluded to as those of Nos. 1 to 8 in a former page. The first six lead to the strait of Malacca; No. 7 is the route for the strait of Sunda, and No. 8 is that taken when bound to either of the eastern channels.

A ship having reached the strait of Sunda, or either of the eastern channels, and proceeding to the northward, must pass West or East of the large island of Borneo. The route then to be taken for China will depend on the prevailing monsoon in the China Sea. Yet, in general, when a vessel is certain of gaining her port with the S.W. monsoon, she should take the strait of Sunda and the straits of Carimata, Gaspar, or Banca. This is the most direct route for reaching China during the S.W. monsoon. But when, on the contrary, the N.E. monsoon prevails in the China Sea, it is necessary, if a ship would not encounter a contrary monsoon, (although she may do so, but it is very trying, and also very tedious,) to take one of the passages East of Borneo, as the strait of Macassar, into the Celebes Sea, from thence, according to circumstances, proceeding northward, and passing East or West of the Philippines. In this monsoon Pitt Channel, to the eastward of Celebes, may also be taken, crossing the Moluccas and entering the Pacific Ocean by Pitt Strait, Dampier Passage, or that of Gilolo, then, keeping to the eastward of the Philippines, the China Sea is entered by the channel South of Formosa.

Thus it may be taken as a general rule that with the favourable S.W. monsoon in the China Sea a ship should go West of Borneo; and with the contrary N.E. monsoon, she should go East of that island.

Lastly, there are two other routes that may yet be taken for China; these are, the Great Eastern Passages, one by Bass Strait, and the other South of Tasmania. These two routes are the safest, when the Cape is passed, towards the month of September, the latitude being run down in the Pacific Ocean, and the China Sea entered by the channels South of Formosa.

Cape of Good Hope to China, with S. W. Monsoon,—Direct Route.—To make the direct passage to China, a vessel should reach the meridian of Amsterdam by the 1st of July at the latest, so as to arrive in China in August. This route, No. 7, as far as the strait of Sunda is already pointed out.

In passing the strait of Sunda, either channel may be taken for the China Sea, passing to it by the strait of

Banca or Gaspar, and afterwards crossing the equator in 105° 30′ or 105° 40′ E.

Inner Route.—The China Sea being entered by the strait of Singapore or Banca, the channel West of the Paracels, (the Inner Route,) will be found the shortest and most direct during March, April, and May; but in June, when the S.W. monsoon is fully established, that by the Macclesfield Bank, called the Outer Route, will be preferable. Some further details may be acceptable concerning this latter route, which should be taken when the S.W. monsoon is well established.

Pulo Aor to Macao in the S. W. Monsoon.—Thus a vessel after passing East of Pulo Aor should make for Pulo Sapata, being careful to allow for the easterly current from the gulf of Siam, that may be expected in the run between these two islands to set her several leagues to the eastward of her reckoning. She should also, on leaving Pulo Aor, run a convenient distance to the northward, to avoid the Charlotte Bank, and not make easting until the latitude of this bank (about 7° 6′) is passed: she should then steer so as to sight Pulo Sapata.

In case of having no observations, when approaching Pulo Condore and that island, great attention must be paid to the lead, as well as when nearing the Catwicks.

After having passed about seven leagues East of Pulo Sapata, a course should then be steered to strike soundings on the eastern part of the Macclesfield Bank, allowance being made for easterly currents, which generally prevail during the S.W. monsoon. This being found, a course should be shaped for the largest of the Ladrone

Islands, which should be passed on the N.E., when the wind is fresh from South or S.W. But at all times it will be better to avoid making the western coast.

A vessel leaving Pulo Sapata towards the middle of September, as soon as she has reached the parallel of 12° or 13°, clear of the banks in the southern part of the China Sea, should keep well to the eastward, as indeed on getting to the northward it is possible she may meet with N.E. or E.N.E. winds: in which case she is again on a bowline, and must make her easting as well as she can.

In the month of October she should make the coast of Luconia, and not steer for the coast of China until she is well North of Cape Bolinao, taking care also to keep a good look out when near the parallel of the Pratas.

A vessel about Pulo Aor, later than the month of September, will find great difficulty in making direct for Canton, on account of the strong southerly currents and the prevailing light winds from the northward. In this case experience has shown that the best route is to the eastward, making the West coast of Luconia.

The passage to the northward may be made at any time of the year along the West coast of Mindoro and Luconia, provided the vessel is a good sailer. It may be as well to keep along the coast, so as to take advantage of the alternate breezes and variable westerly winds which are met with from October to April.

Palawan Route.—A vessel having reached Pulo Aor or Pulo Timoan about the month of October, would adopt the following, called the Palawan Route.

From Pulo Aor she would steer for the South Anambas, and leave these islands to the North, as well as Laut and the Great Natuna she would pass between the Louisa and Royal Charlotte Banks. She would then make for the island of Balambangan, on sighting which she would pass it at a distance of eight or nine leagues, with a southerly wind. But with the wind westerly, she would keep it fifteen leagues from her, and make for the island of Balabac, which she would pass at a distance of nine or ten leagues with favourable winds. With easterly winds she would near these two islands, on account of the strong westerly currents, and then enter the channel formed by the Half Moon, the Royal Captain, and Bombay Banks and the outlying dangers of the northern coast of Palawan. In passing through this channel she would keep generally about thirty miles from the shore, using the greatest caution, in order to avoid all its dangers.

With an easterly wind, when the vessel is to the northward, she would sight the North cape of Palawan and the Calamianes Islands, and then make Luban or Cabras Island, keeping near the coast of Luconia. She will thus reach Cape Bolinao, from which she would keep at a moderate distance, on account of the outlying shoals and the currents which sometimes set into the large bay near it.

A ship having reached Cape Bolinao, will be pretty sure of passing East of the Pratas, unless prevented by a N.E. gale and strong westerly currents, which is frequently the case. It would therefore be prudent to keep along the coast of Luconia as far as Cape Bojador, in order to make sure of sighting the coast of China East of the Lema Isles, from whence Macao would be easily made. Should the season be too far advanced when leaving Balabac Strait for China by the Palawan Route, the Sooloo Islands and sea of Celebes should be crossed, and then the northern point of this island passed into the Pacific Ocean South of Mindanao, a route which we shall shortly describe, forming one of the eastern routes.

A vessel going to the Moluccas, on leaving the strait of Malacca may take the same route: if before September she should make to the eastward after she has entered the Pacific Ocean, and then run down to the Moluccas, either by the Gillolo Channel or Dampier Strait; in September, after she has rounded the North point of Celebes, she may proceed to the Moluccas by the channel of this name.

Eastern Route for China during the N.E. Monsoon.—The Eastern Route for China is taken by vessels from the Cape of Good Hope that reach the meridian of Amsterdam Island between the 15th of September and the beginning of December, and is the route No. 8 of the chart, leading to the eastern channels.

A ship from any port of India to China by the Eastern Route should, if possible, leave in October or November, in order to return by the usual route before the end of the monsoon.

In leaving a port in the northern part of the gulf of Bengal for China, when the time of the season obliges her to take the Eastern Route, she would take the strait of Malacca and the Carimata Channel, or go West of

Sumatra if circumstances required. A ship leaving the same ports at the beginning of the N.E. monsoon, the strait of Malacca and the Palawan Route will be better for her than the long circuit she would be obliged to make in taking the Eastern Route. When leaving the coast of Malabar or Ceylon or any port situated South of the Coromandel coast, she would pass South of Java, and take one of the straits East of this island. The strait of Sunda might also be adopted, and then she might make her easting in the Java Sea and towards the strait of Macassar. On leaving the Malabar coast or any port on the West coast of the Indian Ocean, she would not enter the strait of Malacca.

October and November are considered the two most favourable months for running through the strait of Macassar; this is the first of the Eastern Routes. In other months it appears more advantageous to take Pitt Passage, especially from the middle of December to February.

Vessels reaching the eastern straits in the latter part of January or February, generally take Lombock Strait; in passing it they usually take the channel East of Banditti Island. The channel West of this island may be taken, but it is very narrow. The middle channel, between Lombock and Banditti, might then be preferred, and afterwards the eastern side of the strait kept; from thence the strait of Macassar may be steered for, passing East of Hasting Island and the little island of Pulo Laut; then the coast of Celebes, in order to enter the strait of Macassar.

When, instead of taking Lombock Strait that of Bally

is adopted, with the view of entering the strait of Macassar, northing should be made by passing through the channel between the islands of Pondy and Galion, then round to the West at a good distance the islands and banks of Kalkoon, and then passing the little Pulo Laut on either side as appears best.

A vessel from Allass Strait would steer for Hastings Island, and pass East of it, the same as if coming from Lombock Strait. If from Sapy Strait, in September and October, she would, according to the prevailing winds, pass East or West of the Postillions, then North between Tanakeke and the Tongu Isles; she would then pass at a good distance the islands and banks of Spermonde, N.W. of the bay of Macassar, and would enter this strait, keeping on the coast of Celebes, to pass through. On clearing the strait, when off Cape Dundas, in about March or April, she would cross the sea of Celebes and steer for the eastern end of Bassilan.

A vessel shaping her course for the channel between Bassilan and the West point of Mindanao, must be careful to keep well to the eastward if the wind permits, in order that she may not be drifted by the westerly currents among the Sooloo Islands. If she finds herself to leeward of them, there are good channels between the islands situated West of the Sooloos; she would then cross the sea of Mindoro, keeping near the coast of the Philippines (Mindanao, Negros, Panay, Mindoro, and Luzon). At the entrance of the channel, between Mindanao and Negros, as well as between Panay and Mindoro, strong N.E. winds and westerly currents are generally found. She would, therefore, guard against them when running from one island to another, so as not to be set to leeward.

If a ship leaves Bassilan Strait with steady S.W. or southerly winds, she may steer direct for Point Naso, or a little East of its meridian: if the wind is changeable and uncertain, she would keep along Mindanao till Point Galera is reached, and then cross to Point Naso, endeavouring to keep near Negro Island in crossing from one point to the other.

From Point Naso she would steer North along the West coast of Panay, taking every precaution against the dangers lying off this coast, she would then pass the islands off the S.W. point of Mindoro, and enter the channel either East or West of them and the shore of Apo. To enter the East channel, in an easterly wind, she would keep at two or three leagues from the coast of Mindoro; but with a westerly wind, taking care not to go more than nine or ten miles from the coast until she is North of the Apo Banks, she would thus clear the strait of Mindoro, and after having doubled the promontory of Calavite, and passed Luban and Goat Island, she would follow the coast of Luzon as far as Cape Bolinao. When she has reached this cape, she may be pretty certain of passing East of the Pratas, and reaching Macao. However, it is prudent to steer North as far as Cape Bojador before crossing for the coast of China. At this season, also, when the sea of Celebes is reached, a vessel may enter the Pacific Ocean by passing South of Mindanao. For which purpose, if the wind permit, she will steer direct for the islands of Serangani, and afterwards pass between these islands and Mindanao, or else South of the former. From thence she would pass between the Meaugis and Tulour Isles, in order to double the North cape of Morty Island, with the wind at N.E. Should any difficulty arise in taking this route, the channel formed by the islands of Tulour and Saugor would be adopted.

Pitt Channel, which leads, as already observed, into the Pacific Ocean, by either Gilolo or Dampier Channel, is preferable to the Strait of Macassar during December, January, and February. A vessel, at this season, arriving at the strait of Sunda, on her way from Bengal, or at the eastern straits, on her way from the cape of Good Hope, would adopt this channel when bound to China; this it is that forms the Second Eastern Route.

When a ship, as it frequently happens, from the gulf of Bengal takes the strait of Sunda instead of passing along the South coast of Java, on entering the Java Sea she should pass North of the Thousand Isles, and then steer eastward, leaving the Watcher Isle to the North, on her way to the strait of Salayer. In case of touching at Batavia: on leaving this port, after having passed Edam Island, she would steer so as to leave Burakin Island to the northward, and, having passed it, would steer for Salayer Strait. The best course through this strait with a N.W. wind is to pass South of Mansfield Bank. At night, or when the wind is not steady, it is better to keep to the northward of it, along the coast of Celebes. From the strait of Salayer she would make for Bouton Strait, or, what would be still better, if the wind is West, she would pass South of this island, keeping the S.E. point well on board, with the view of avoiding the rocks off it to the southward of the islands of Toukan Bessy. She would then pass along the eastern coast of Bouton Island, and, having reached the N.E. end of it, if the wind is fresh from N.W., she would make to the northward for the island of Waigiou, and from thence for the island of Kulla Bessey. This is a necessary precaution for dull sailing vessels in December and the beginning of January, because the wind becomes variable about this period, and veers to N.N.W., producing a strong southerly current. The winds and currents are, however, very variable in Pitt Channel, and it may be crossed almost anywhere. It is prudent, however, to keep the weather shore when northerly winds prevail.

A vessel falling to leeward of the N.W. point of the island of Bouro should make every exertion to pass it quickly. To do this, instead of working to windward, it is better to run to southward of the island and pass to the eastward of it into Pitt Strait. Vessels which leave Amboyna in the N.W. monsoon make to the northward along the East coast of Boero, where the wind is variable and off the land; while beyond Manipa and in the channel which separates it from Ceram, in this season southerly currents prevail. A vessel having reached Pitt Passage by the foregoing routes will be guided by the directions hereafter given.

A vessel intending to use Pitt Channel should take either the strait of Bally, Lombock, Allass, or Sapy, and on leaving either of these straits would make for that of Salayer, crossing the eastern part of the Java Sea, and afterwards steer for Pitt Channel. For a vessel from the Cape the strait of Ombay is preferred, as being the most direct and larger than those further West, and the wind there being generally less variable.

A vessel making for Ombay Strait must pass either North or South of Sandal-Wood (Sumbawa) Island; but it is better to pass South of it and then between Ombay and Timor, and, having rounded the eastern end of the former, she would then steer North, keeping ing to windward, so as to pass West of Boero Island. If this cannot be done, she would pass East of this island, between it and Manipa, and then take Pitt Channel. Having entered Pitt Channel, she would then steer East, passing between Kulla Bessy and Boero; and from thence, if no current be found, she would steer direct through Pitt Strait. If the current sets northward, she would keep from the islands bounding the northern side of the strait.

When near the meridian of the East point of Oby Major, and intending to take Dampier Strait, a vessel should continue to run eastward. This strait seems favourable for good sailing vessels, especially in January and February, when N.E. winds are getting more easterly. In March, when the N.E. winds become weak, the strait of Gilolo appears the best for entering the Pacific Ocean. This latter is also wider, and a ship may work both night and day in it, and, besides, the currents there are seldom strong. On leaving Pitt and Dampier Straits great caution should be used to prevent being set on the North coast of New Guinea, and a vessel should therefore manage so as to round Point Pigot close, looking out at the same time for Buccleugh Bank, to the East of the eastern coast of Waigiou.

Pitt Strait.—Pitt Strait should only be adopted when it is compulsory to do so. In this case a vessel should keep mid-channel, to avoid being set to either shore by the tides, and should therefore make short boards, not approaching either shore. On reaching Jackson Island, she would pass about five miles North of it: and, having passed the reef off the extreme of the island of Batenta, she would steer North for Point Pigot. Horsburgh advises ships to avoid this strait by all means.

Dampier Strait.—A vessel entering Dampier Strait, on passing the meridian of the East point of Oby Major should steer East, and go between Pulo Popo and the Kanary Isles. Sometimes vessels pass between the Boo Islands and Pulo Popo; this is an advantageous route in N.W. winds. She would then run on for Fisher Island or Cape Mabo, and from thence pass between Pigeon and Foul Islands, looking well out for the dangers on the North shore of Dampier Strait; and on leaving it she would keep closer to Pigeon than to Foul Island, and she would steer so as to sight Pigot Point, in order not to be set down to the coast of New Guinea by the northerly swell which prevails out at sea. She would also take care to avoid the Buccleugh Bank.

The tides in Dampier Strait are very strong and the currents very irregular, varying at a rate from one to five miles an hour. In the height of the N.W. monsoon, in the narrow part of the strait, between Pigeon and Foul Islands, the ebb at the time of syzygy runs between four and five miles to the E.N.E. for six or eight hours, and between one and three miles in neaps. The flood sets S.W. for three or four hours, but is weak. During the height of the S.E. monsoon, in this part the flood sets West for eight or ten consecutive hours, turning successively W.S.W., S.W. and S.W.b.S.: it then attains its greatest velocity, which at springs sometimes exceeds five miles an hour, and at neaps four miles. The ebb at this season runs E.N.E. or N.E., and is neither strong nor of long duration.

A vessel, on leaving Dampier Strait, having entered the Pacific Ocean, should make easting rapidly, keeping in a low latitude, or between the parallels of 1° 30′ and 3° N., which she will do easily, sometimes even in December and January. In this track an easterly current is found. She will thus make good her easting as conveniently to pass either East or West of the Pelew Islands, which will depend on the qualities of the vessel and the strength of the N.E. monsoon. In all cases she would not go far East, on account of the islands of Goulou and Guap, near which, in November and December, heavy squalls from the westward are met. From the Pelew Islands she would steer for the Bashee Islands, allowing for the westerly currents, which set about twelve or fifteen miles a day. From December to the middle of February it would be best to pass East of the Pelew Islands.

A vessel leaving Dampier Strait towards the end of the N.E. monsoon, would not go so far East into the Pacific. At the end of February and March she would pass West of the Pelew Islands, as the winds in these months are variable and shift to E.N.E. When she has reached the North part of Lucon she would enter the China Sea by either of the great routes between Formosa and Lucon. However, with N.E. winds and at the beginning of the monsoon, it will be necessary to pass North of the Bashee Islands, and either North or South of the Cambrian Rock. She will thus approach the South point of Formosa, and must pass, if during the daytime and the weather is fine, between this point and the Vele Rete Rock. During the night, or should bad weather prevent her taking this route, she would pass North of the Bashee Islands, keeping close to them.

Whichever may be the channel by which the China Sea is entered, a course should be adopted to sight, if possible, Pedro Blanco, and enter the Canton River by the Lema Channel.

Gilolo Strait.—The strait of Gilolo is divided into two parts by Gebe Island, and the channel between this island and Gilolo takes the name of Gilolo Strait. That between Geby and Waigiou has been called the strait of Bougainville, that officer having passed through it in 1772.

All the channels leading from Pitt Passage to the strait of Gilolo are free from danger. In the N.W. monsoon, however, that between Pulo Gass and Kakik Island is preferred, as being the largest; for the other, between Pulo Pisang and the Boo Isles, is at this season too much to leeward. A vessel entering Gilolo Strait, passing, as above said, between Pulo Gass and Kakik, should round the southern point of the first of these islands closely, in order that she may not miss the channel by the drift of the easterly current, which often prevails there. When she has passed East or West of Pulo Gass, according to the channel taken, she should continue on between Cape Tabo and Geby Island, and if by night, should give the Fairway Bank and Widda Islands a good berth. Should the wind be light she should also keep as close to the islands on the West coast of the strait as possible, on account of the N.E. and easterly currents.

Should the wind be foul, no time should be lost in trying to pass North of Geby; but it should be left to the northward, the ship passing between it and Gagy, and entering the Pacific by one of the channels near Syang. But, whenever it can be done, the West channels between the coast of Gilolo and the Shaupee Isles, or between these islands and Syang, is preferable, because with a northerly wind a ship can pass to windward of the islands of Ayou and Asia. Should there be any difficulty in passing West of the Asia Isles, the channel formed by them and Ayou, or even that between this latter and the North coast of Waigiou may be adopted.

Having gained the Pacific, the vessel should endeavour, as soon as possible, to attain and keep between the parallels of 1° 30′ and 3° N., and make her easting there, as southerly and S.E. currents are found there; and she will thus attain the latter part of the route, previously mentioned, from Dampier Strait to the China Sea.

Great Eastern Route from the Cape to China in the N.E. Monsoon.—The following, called the Great Eastern route, may also be adopted during this season when the Cape is left in September.

From South of the Cape a vessel should steer East, keeping between the parallels of 38° and 40°, or thereabouts, as far as the meridian of Cape Leuwin. From thence one of two routes may be taken: that South of Tasmania, or that by Bass Straits.

The first route was adopted by Captain Butter of the ship *Walpole.* He left the Cape of Good Hope in the end of September; on the 31st of October he sighted the S.W. point of Tasmania; on the 18th of November he sighted the island of Anno Bom, one of the New Hebrides; he left these islands a little to the West, crossed the equator West of the island of Paanopa, in about 161° 40′ E. long., and crossed the archipelago of the Carolines. On the 21st of December he passed the Marianne Islands; on the 30th he entered the strait South of Formosa, and on the 1st of January anchored in the harbour of Macao.

Thus the voyage lasted three months; that is, it occupied only a little more than the time that is generally occupied by the direct route to China during the favourable season.

Thus about two months is the time from the Cape to the strait of Sunda, and three weeks from the strait of Sunda to the Canton River.

Again, instead of passing South of Tasmania, ships generally take Bass Straits. This route was taken by the *Athenian* in 1804. On the 11th of October she passed Amsterdam Island; entered Bass Straits the 28th of October; passed West of New Caledonia and the

Hebrides, and then between these latter and the Solomon Isles. She crossed the line in 160° E. long., and sighted the coasts of China on the 28th of December. Notwithstanding this, and although the route is much shorter, Admiral Krusenstern prefers passing South of Tasmania. His opinion is founded on meeting westerly winds in high latitudes, and by passing South of Tasmania northerly winds and southerly currents, often met at the entrance of Bass Strait, are avoided.

CHAPTER XIII.

RETURN ROUTES BETWEEN CHINA, THE PHILIPPINE ISLANDS, AND INDIA.

In the homeward routes from China to Europe, the chief thing to do is to gain the S.E. Trade as soon as possible, with which a vessel will soon reach the Cape.

A ship from China to India should endeavour to reach the strait of Singapore as soon as possible, and then run through the strait of Malacca. Thus in the two monsoons: When a ship leaves China in the N.E. monsoon for Europe or India, she would make for the strait of Banca and Gaspar, or for that of Singapore. In March and April the quickest route is the outer one by the Macclesfield Bank, and in these two months she would keep out to sea as far as the latitude of Pulo Sapata.

In the other months, on the contrary, a vessel would take the inner channel, between the islands of Hainan and the Paracels, and would thus without difficulty reach the straits of Singapore, Gaspar, and Banca. From the two latter a course should be steered for the strait of Sunda. On leaving this strait the parallel of 10° N. should be crossed in 100° E. long., and then a direct course shaped for the South point of Madagascar. The ship would then follow one of the routes given in the latter part of chapter ix. This route crosses the very middle of the course of hurricanes; consequently, they are frequently met by vessels from the eastern seas.

Inner Route from Macao to Pulo Aor.—The inner route is the most direct that can be adopted for reaching the straits leading from the China Sea; it also has the advantage that vessels have the wind aft as soon as they have passed the Great Ladrone.

A ship taking this route should steer from the Great Ladrone so as to pass near the Taya Isles and the Paracels at a convenient distance to the West. The current may be estimated as setting westward at the rate of fifteen or twenty miles a day, for currents are strong near the coast of China, although not so perhaps out at sea. Should it be found that the ship is drifted much to the westward, she must shape her course so as to allow for it until she has reached the parallel of 17° N., and entered the channel between the Paracels and Cochin China. Having reached this parallel, and the meridian of 109½°, a course should be steered so as to sight Cape Varela or the Pagoda.

With clear weather and an E.N.E. or N.E. wind, she may sight Pulo Canton, (called also Collao Kay,) or the coast to the southward of this island, and then keep the shore at a moderate distance; but if the weather be cloudy, and the wind has a tendency to become easterly, it will be more prudent not to approach the coast till she is in the latitude of Cape Varela, nor enter the bay of Phuyen to the northward of this cape. Should the conical mountain on the North shore of this bay be visible, it will indicate the position of the cape, for, as night approaches, the pagoda on the height, which commands it, is concealed by clouds. Having passed South of the parallel of 15° N., it will be found near the land that the current sets southward; for between 14° 30′ and 11° 30′ it often sets at the rate of forty, fifty, and even sixty miles a day; but it is most uncertain.

When land has not been seen to the northward of Cape Varela, it is indispensable to make this cape, from whence the coast may be kept at a distance of twelve or fifteen miles.

When off Cape Varela, distant about four or five miles, a vessel may steer along the shore by day, but at night must be careful to avoid Pyramid Isle, and those near to it. If the night be fine, she can sight Pyramid Isle and its neighbours, for in a clear night it may be made out at a few miles distant. Water Islands should then be steered for, about twenty-one miles further South. Having reached these islands, if the land appear further off than four leagues, it will be necessary to approach it in order to sight the mountain of *False* Cape Varela, which may be distinguished amidst the high land of the coast by its elevation and gentle slope towards the sea.

A vessel desiring to keep in shore and pass West of the Dutchman Bank, should cross Padaran Bay as soon as she is abreast of the high lands of Cape Varela. This is necessary, because in this part the currents take a S.S.E. direction, and it is very difficult to approach the coast against them. In a good position, when crossing the bay, soundings are found with 39 to 50 fathoms. Then at night Cape Padaran should be made on the starboard bow, in doing which great care must be taken, as it is most difficult to distinguish it in the distant land at the bottom of the bay.

Cape Padaran being once sighted, may be passed at a distance of three to six miles, and a course steered then so as to pass Pulo Cecir at the same distance from land, leaving it to the West. Should the Cape be only a mile or two off, a course should be steered so as to pass at a convenient distance from Pulo Cecir. On having passed this island in daylight at a distance of four to six miles to seaward, it should be brought to bear N.N.E. ½ N. before it is lost sight of from the deck; then steering W.S.W. and S.W.b.W. for six or seven leagues, according as most convenient, will pass West of Dutchman Bank, when a South course may be steered for Pulo Aor.

This route is not dangerous when the night is sufficiently clear to admit of the Cavern or Tron of Padaran being made out. In this case, being three to five miles from the cape, a course may be steered until the Cavern is made out, and when it bears N.b.E., the vessel will be off Pulo Cecir. In this case if soundings are found at ten or eleven fathoms, she should stand off from it, for the island is too low to admit of being seen easily at night, and the soundings are too irregular to be of service. The Cavern bearing N. 16° E., Pulo Cecir will be in the same direction, and by running six or seven leagues between W.S.W. and S.W.b.W., will lead West of the Dutchman Bank, when Pulo Aor may be steered for. Should the night become dark when near Cape Padaran, and the land not distinguished about the Cavern, the vessel should be kept between South and West, to about twelve or thirteen leagues from the cape, and the coast should not be approached to less than eleven fathoms, nor the Holland Bank in less than twenty fathoms.

Between the western edge of this bank and the eastern one of the Britto Shoal, which is the nearest to it, is a distance of fourteen or fifteen leagues; which channel may be taken during the night. A vessel should keep in soundings of fourteen or fifteen fathoms until she is five or six leagues South of Pulo Cecir; and when about thirteen leagues S.W. of Cape Padaran, it will be better to run on to the southward and westward for two or three leagues more, so as to give a good berth to the Holland Bank. A vessel should not have more than twenty fathoms until this bank is passed, nor less than fifteen fathoms when near the Britto Banks; by keeping a little to the westward when passing between these two shores, soundings varying about nineteen or twenty fathoms will be found. After passing the western edge of the Holland Bank, a vessel should keep in twenty or twenty-one fathoms, steering towards Pulo Aor.

The route between Pulo Cecir de Terre and the Holland Bank can only be taken in the night by commanders who are well acquainted with these parts; consequently, while waiting for daylight, a vessel is often obliged to lay by off Cape Varela. Besides the loss of time resulting from this, a ship has to contend with a heavy sea when the breeze is strong; for which reason, as already observed, the passage outside Pulo Cecir and Pulo Sapata is to be preferred.

Route for Passing outside Pulo Sapata.—A vessel finding herself at nightfall near Cape Varela, with bad weather or a wind too strong to haul up to, and not wishing to pass between the Holland Bank and Pulo Cecir de Terre at night, should steer a course to pass to the eastward of Pulo Cecir de Mer, and afterwards Pulo Sapata on the following morning. She may even run far enough out to sea, if the weather is cloudy, so as to pass a good distance outside of these islands. Indeed, when the wind is strong, the current sets S.W. and W.S.W. very strongly, and sometimes towards Pulo Sapata. A vessel would then be obliged to pass the night in the narrow channel between this island and the Little Catwick.

In daytime, with fine weather, a vessel may keep as near as she likes to Pulo Cecir, and pass between Pulo Sapata and the Large Catwick. She may even pass between the two Catwicks, but must remember that the Paix Rock is in their neighbourhood. Thence she might steer direct for Pulo Aor, and, reaching the neighbourhood of Pulo Timoan, in cloudy weather should keep in soundings of thirty-one fathoms, so as to pass East of this island for Pulo Aor. As these islands are often concealed by fog, great care must be taken to avoid them and attend to the reckoning, especially during the night.

Near the Anambas, and to the northward of them, a vessel has generally thirty-five to forty-two fathoms. Between the parallels of $5\frac{1}{2}°$ and $5°$ N., these depths decrease in the western part of the channel, and twenty-seven fathoms are found on the meridian of Pulo Timoan.

Route from Pulo Aor.—When a vessel has passed East of Pulo Aor, at a distance of two to four leagues, for the strait of Banca, she would steer to the eastward of South, according to the wind and prevailing currents, and pass outside the Geldria Bank, which may be avoided by keeping in a depth of twenty fathoms, and when between the parallels of 0° 56′ and 0° 40′ N., a course should be steered so as to cross the equator in seventeen fathoms, and pass four or five leagues from the East point of Lingin, if the current will admit. In all cases a vessel should guard against westerly currents, which are sometimes found in these parts, taking care that she is not set over to the Ilchester Shoal, South of this point.

Route to Banca Strait.—When off the East point of Lingin, about five leagues distant, a vessel would pass between Pulo Taga and the Seven Islands (or Toojoo.) If she is further to the eastward she will steer more westerly, and, having cleared the channel between Pulo Taga and the Seven Isles, she would make for Point Batacaran, the western extremity of Banca Strait, keeping at a distance of six and a half or seven leagues from it, in order to avoid the Frederick Henry Rock.

Outer Route from Macao to Pulo Aor.—When the outer channel is adopted from Macao for Pulo Aor, the vessel would pass at a short distance West of the Ladrones and neighbouring isles. On leaving the Great Ladrone, as strong winds and a heavy sea, with strong currents are found, a vessel should steer to eastward of South for the Macclesfield Bank, and in moderate weather she would endeavour to reach the East part of it. When at twenty leagues East of the meridian of the Great Ladrone, if there be difficulty in obtaining soundings, a vessel may then consider herself East of the Macclesfield Bank.

A vessel having adopted the outer route in November and December, with strong winds and no observations for several days, should strike soundings on the Macclesfield Bank; but if certain of her position, they may be neglected. The bank being very wide, East and West, and the soundings very irregular, the depth can only be an uncertain guide as to her real position.

On leaving the Macclesfield Bank she would steer for Pulo Sapata. If a vessel should have had soundings on that bank, and is on the same parallel, she should shape her course for that of Pulo Sapata; and should she not sight this island, she would steer West, so as to obtain soundings at thirty to thirty-five fathoms.

In thick weather vessels uncertain of their position should not steer direct for Pulo Sapata, or attempt to round the island at night, as it is difficult to distinguish. But as a general rule, they should keep well to the eastward of it until on the parallel of 10° N., and by standing W.b.S. obtain soundings. Some captains, on leaving the Macclesfield Bank in March, April or May, run as far as the latitude of Pulo Sapata, keeping well off to the eastward of the island. In adopting this route, however, care must be taken to allow for the S.E. currents, which might set a ship on the banks to the E.N.E.

and East of Pulo Sapata. When the parallel of 10° N. is reached, we should steer between West and South until soundings are found in 28 fathoms; a course should then be steered for Pulo Aor or Pulo Timoan. If bound to the strait of Singapore, when in lat. 7° 6′ N., to avoid the Charlotte Bank, the soundings should not be more than 23 or 25 fathoms.

In March and April vessels returning to Europe should keep well to the eastward, to pass between the Natunas and Anambas Islands, and take the strait of Gaspar.

China to the Gulf of Bengal in the S.W. Monsoon.—It has long been considered impossible to cross the China Sea from North to South against the S.W. monsoon. With a good ship, however, this may be done in any season by adopting the inner route West of the Paracels. By leaving the Great Ladrone, with one of those breezes from East and S.E., which are very frequent there, and last for several days, a vessel would probably reach the strait of Singapore if bound to Bengal sooner than by taking the eastern routes by the coast of Luconia. In this case she should pass at a moderate distance from Hainan and the coast of Cochin China, but not near enough to feel the effect of the strong N.N.E. currents of that coast, nor so far out to sea as to lose the advantage of the changes of wind and smooth sea. Besides which, out at sea the wind is frequently South and S.S.W.

The route just pointed out applies only to ships bound to the gulf of Bengal by the strait of Malacca; but in some years it is attended with much difficulty. Vessels for the West coast of India, or the Cape of Good Hope, should not adopt it if they meet with contrary winds at the entrance of the strait of Malacca, because they will find great difficulty in getting through, and the winds will continue contrary for a space of 12° or 14° of latitude after having passed Cape Acheen, that is, up to the equatorial limit of the Trade winds. In May and June a vessel should not make to the southward for the straits of Gaspar, Banca, or Carimata.

A ship leaving Canton with the S.W. monsoon for Europe, or the eastern coast of India, would find it prudent to adopt one of the following routes, according to the time at which she leaves China, while the S.W. monsoon prevails.

First Eastern Route.—A ship leaving Macao, or the Great Ladrone, in the end of April or beginning of May for the first Eastern Route, that is, the Mindoro Strait, should make to the South as far as the Macclesfield Bank, if the wind permit, so as to reach the N.W. extreme of Mindoro without tacking in case of the wind shifting to S.W. From near the Macclesfield Bank she would stand S.E., keeping her wind if it is at all to the S.W., and should it not admit of her weathering the point of Calavite she should work along the coast of Luconia with the variable winds, with which she will reach the N.W. extremity of Mindoro.

The channel East of the bank of Apo should be adopted for crossing the Mindoro Strait, giving the Mindoro coast a berth of some miles, if the wind is variable; a distance of nine or ten miles is necessary if the S.W. wind is steady; she will then pass the islands of Ambolon and Ilin at a distance of about fifteen miles.

Should the wind admit, a vessel may cross the strait of Mindoro, passing West of the Apo Bank, in the Northumberland Channel, formed by this bank and the Calamianes. She would then keep along the coast of Panay, working, if necessary, at some distance from this island, according to circumstances, and would near the island of Quinilaban, so as to pass the dry sand bank between this island and the coast of Panay.

Having reached Cape Naso, she would then stand for the strait of Basilan, making it well to the southward and westward, when the wind is from these quarters; but steering direct for it if the wind is easterly. The S.W. extreme of Mindanao being gained, it will be better to take the strait of Basilan than those formed by the islands to the S.W., the former route being the shortest; the Celebes Sea will thus be entered, and the ship will make for the strait of Macassar. Instead of persevering in working at the entrance of the strait of Basilan against S.E. winds, Captain Spratly recommends steering West, in order to pass West of the archipelago of the Sooloo Islands, between the point of Nusang and the island of Tawee-Tawee. There are two little islands close off the S.W. point of this island, bearing S.W., near the island of Sibbootoo, and forming a good channel leading direct to the Celebes Sea. This channel is safe, and Captain Spratly says, it is easy of navigation both by night and day. He considers four hours sufficient for passing from one sea to the other by it, while under similar circumstances it has sometimes occupied four days in going from one sea to the other by the strait of Basilan.

To leave the Celebes Sea, a vessel may either take the Macassar Strait or the Molucca Channel. Some navigators prefer the latter when the S.E. monsoon prevails North of the equator. In fact, it is difficult, without a tedious working to windward, to reach Allass Strait from the strait of Macassar; while by taking the Molucca Channel the S.E. monsoon is found in a latitude sufficiently to the eastward to enable one to take whichever Eastern Channel is preferred. But vessels bound to Batavia, or the strait of Sunda, will find it best to take the strait of Macassar.

On leaving the strait of Basilan, if the easterly wind is well established, a vessel should steer so as to make Cape Donda to the S.S.E. or South; but most generally, from the winds veering westward near the northern entrance of the strait, and the current setting eastward, it is prudent to keep as much as possible to the westward, in order to sight Point Kanneeunegan. According to Captain Spratly, a ship off Cape Rivers is sometimes set to the eastward by the current along the coast of Celebes, and after useless contest with it, is sometimes obliged to take the Molucca Channel.

A ship having entered the strait of Macassar, should keep along the West coast of Celebes, passing East of the Little Paternosters, being very cautious, on account of the dangers North of the islands of Nossa Seras, in passing between them and the Great Pulo Laut. From thence she should steer for the strait of Allass, or one of the straits leading into the Indian Ocean. If bound to

Batavia or the strait of Sunda from the strait of Macassar, she should steer South if the wind will permit, and pass North of the Little Paternosters for the coast of Borneo, keeping along this coast and providing against the dangers off it, as well inshore as to seaward. She would then enter the Java Sea, and reach Batavia or the strait of Sunda without difficulty; and thence the Indian Ocean, and make for the Cape, or the western coast of India, by the routes already described.

A ship taking this route, and meeting with contrary winds from the strait of Baselan, so as to be unable to reach the strait of Macassar, may take the Molucca Channel, and should then steer for the islands near the N.E. end of Celebes; and passing between the islands of Banka and Bejaren, would round the N.E. point of that island, and then steer to the southward, through the channel formed by Lissa Matula and Oby Major, which is the most frequented; or, if the wind should not permit her reaching it, should take the Greyhound Channel, between the islands of Albion and Hammond (West of Xulla Tallyabo.)

When it is difficult to get to the southward in the Molucca Channel, dull sailing vessels might try to do so by keeping near the West coast of Gilolo; thence they might enter the strait of Patientia, between Gilolo and Batchian, or the strait of Batchian, formed by the island of this name and Tawally and Maregoland.

However, a ship having reached the northern extremity of Gilolo or Morty in the height of the S.W. monsoon, should rather pass through the Gilolo Channel than that of the Moluccas, because it leads more directly to Pitt Channel, by which she can gain the eastern straits.

On leaving the Molucca Channel the strait of Ombay may be adopted if desirable. A ship should then pass close to Oby Major, in order easily to round the East coast of Bouron, and so pass between this island and that of Manipa. She would then run to the southward into the Banda Sea, where the winds are generally from E.S.E.; on leaving Manipa she would endeavour to pass to the East of Ombay, and having crossed the channel formed by this island and Wetta, would follow the West coast of Timor, and enter the Indian Ocean between Semao and Savu. This is the shortest route during this season from Pitt Passage to the Indian Ocean.

Second Eastern Route.—The Second Eastern Route for the Cape or West coast of India from China, with the S.W. monsoon, is adopted from the middle of May to the end of July. This route is by taking the Pacific Ocean East of the Philippines, and passing through Pitt Channel. In August it is too late to adopt this route, and a ship obliged to leave Macao then, should follow the coasts of Cochin China and Cambodia, unless from being a bad sailer it may be better to defer her departure until September.

With southerly or S.W. winds, a ship to pass East of the Philippines should leave Macao by the Lema Channel, and then steer South in order to enter the Pacific Ocean without tacking. If the wind admits, the channel between the Bashee and Babuyanes should be adopted. Having reached the Pacific Ocean, S.W. winds at this season will generally be found, with easterly or N.E. currents; she should then steer S.E. in order to avoid Cape Engano and Luconia, tacking if necessary so as to pass neither too far out nor too close, and taking care not to round the Pelew Islands further to the eastward than is necessary.

The best route for making southing is then East of the isles of St. Andrew, Current, Marriere, Lord North, and the dangerous Helena Shoal. If easterly currents are met they will not be strong as far as the Pelew Islands; but between lat. 5° and 2° N. they set at the rate of thirty to sixty miles per day. This part must therefore be crossed as quickly as possible if the wind is West, as it frequently is; and if the wind is light, a ship may be set far to the eastward by this current. But from the lat. of 2° N. to the equator, a westerly current will be found, while near Dampier Strait it is again running to the eastward.

Having rounded to the eastward the island of St. Andrew, a ship should endeavour to keep between the meridians of 132° and 133° E., and when in 1° N. lat., if Dampier Strait is to be taken, she should make for Point Pigot.

The Strait of Gilolo.—The strait of Gilolo being larger than that of Dampier, is often preferred for that reason, and it has few difficulties to overcome in reaching Pitt Passage.

When Gilolo Strait is to be adopted, on leaving the parallel of 2° N. a ship should steer for the isles of Asia, and round them on the North if the wind permits, unless she passes between these islands and Ayou.

Having passed the islands of Eye and Syang, she

would then go North or South of the island of Geby, and if the weather be not favourable, instead of the strait of Bougainville she might take that of Gilolo, which is North of it; and in crossing this strait she should keep near the eastern coast, and enter Pitt Channel between Pulo Pisang and the Boo Isles, or else, according to circumstances, between Kekek and Pulo Gass.

Dampier Strait.—A vessel entering Dampier Strait should round Point Pigot at a distance of six to twelve miles, and then steer for King William Island, keeping it West of her; when about nine miles from it, she should steer for Pigeon Island, and pass two or three miles South of it; she may then cross the strait, taking care to avoid any dangers in her way.

On leaving Dampier Strait she would go close round Cape Mabo, so as if possible to pass South of Pulo Popa; or she may pass North of this island and enter Pitt Channel between the Boo Islands and Pulo Popa. In Pitt Channel she should keep mid-channel, borrowing rather on the southern than on the northern side. Having reached West of Pulo Popa, and cleared Pitt Passage, passing between Ceram and Bouron, the Indian Ocean may be entered by the strait of Ombay or one of those westward of it.

Ombay Strait.—The strait of Ombay is the most direct route to the Indian Ocean in the S.E. monsoon; a vessel would then follow the last route in chapter ix. p. 180. if bound to Europe. If intending to take the strait of Salayer, or those of Allas or Sapie, the N.W. part of Bouron should be gained, and thence the most northerly of the Toukan Bessy Group should be rounded at two or three miles distance; and from thence enter the strait of Salayer.

Route for the Gulf of Bengal in the S.E. Monsoon.—On leaving this last strait for Bengal during the S.E. monsoon, a ship should pass near the large Solombo, leaving the Brill Bank either to the South or North, and from this island should pass round the banks situated off Pulo Mancap, and cross the Carimata Channel, then steering for the North extremity of Banca take the strait of Dryon or Singapore, for Malacca Strait. It is likely that by this route she would speedily reach the gulf of Bengal.

The straits of Sapie and Allas are generally taken by ships bound to the Cape or West coast of India. To enter Sapie Strait after rounding Salayer Isles to the South, the North part of Comodo should be made, and then the West channel between Gilibanta and Goonong Apee.

If Allas Strait be adopted, Plate Island should be sighted, from which island a course may be steered for the entrance of the strait, the vessel keeping the chain of low islands, N.W. of Sumbawa, at a respectful distance; after which she may cross the strait. She may then follow the route given in chap. ix. to the Cape of Good Hope, or those in chap. xi. if bound to the West coast of India.

Navigation of the China Sea during the Monsoons.—The passage to the northward or southward against either of the monsoons, is continually made by clippers, vessels which carry opium to China. The following observations are those of Commander P. J. Blake, of H.M.S. *Larne*. To his own private remarks this officer has very properly added those communicated to him by captains of clippers well acquainted with this navigation.

As a general rule, vessels working up against the monsoon, when it is at its height, or when it is favourable, whether they are going North or South in the China Sea, should, if near them, pass to leeward of the Paracel Islands and the shoals off them, and also to leeward of the shoals of Pratas and Scarborough, a precaution extremely necessary on account of the currents which constantly and invariably set towards these shoals.

There is, however, one exception to this rule, and it is when working to the northward against the N.E. monsoon, and approaching the parallel of 14° N. In that part there is a large space of open sea, and a vessel should make to the eastward as far as the coast of Luconia, so as to be able to lay up for Macao, and reach it on the other tack.

Macao for Singapore.—In coming down the China Sea with the N.E. monsoon from Macao for the straits of Singapore, Banca, or Gaspar, the direct course is the one most generally adopted. The Paracels are generally left to the eastward, at no great distance, and in general a southerly current is found near Hainan, running from thirty to fifty miles in twenty-four hours. Between the parallels of 14° and 11° currents have been found running sixty miles per day. After passing the Paracels, the coast of China may be gradually approached in the latitude of Cape Varela or the Pagoda. From thence the course should lead thirty or forty miles East of Pulo Sapata, and from this island to Singapore the route is direct, being West forty miles to the Anambas, always making Pulo Aor if possible, and to verify the position of the vessel soundings should be frequently obtained. From Pulo Aor one of the straits abovementioned may be easily reached.

Carimata Channel.—The route generally taken from Pulo Aor to the straits East of Java, during the height of the monsoon, is as follows. A vessel being about five leagues East of Pulo Aor, should steer so as to sight Victory Island, from whence, steering West, she would make St. Julien, St. Esprit, and St. Barbe, at a good distance. On leaving this latter she should steer for Sourouton and Carimata, if this channel is to be adopted, which at this season is the surest route for reaching the eastern straits. From Sourouton a course should be steered for Pulo Mancap, the Java Sea then entered, keeping a good distance to the West and South of the dangers off this island. The Carimon Java should then be steered for, and then the North coast of Java approached, running eastward along this coast and that of Madura. Isle Pondy would then be passed on the East, from thence, avoiding the Minden Rocks, a course should be shaped for Cape Sedano, at the North entrance of Bally Strait, and the strait entered by passing either East or West of Gilboan.

Some seamen prefer Lombock to Bally Strait, which is further to the eastward and is larger.

A ship leaving the China Sea for the strait of Gaspar, will generally take the West or Macclesfield Channel.

In this case, from Pulo Aor she would steer for Pulo Loty, passing fifteen or eighteen miles East of it if the wind veers at all to the N.E. From Pulo Loty she would make for Gaspar Island, passing on either side of it, and at the distance of about two miles round the West side of Pulo Leat, leaving the Vansittart Shoal and the rock of Dordelen to the eastward, in the southern part of the strait, and then enter the Java Sea for the eastern straits.

The preceding instructions apply chiefly to vessels leaving China at the beginning of the N.E. monsoon, and for leaving the China Sea the strait of Banca is adopted in preference to the preceding ones. In this case, Pulo Taya should be made and passed to the West, making afterwards for the point of Batacaran, and the peak of Monopin of Banca, at the northern entrance of this strait.

At the end of the N.E. monsoon it will be better instead of sighting Pulo Aor to make the North Natunas, because at this season S.S.W. winds are found, varying to S.E. and East. If, therefore, a vessel desires to reach the strait of Gaspar from the North Natunas she should pass fifteen or sixteen miles West of the Great Natuna and West of the Haycock; then round the Victory and Barren Islands on the East, and on the West those of Camel, St. Julien, and St. Esprit, if the wind permits; or, if not, whichever channel between these islands that will suit best. From the St. Esprit Group she would steer for St. Barbe, and pass about nine miles West of it: then for Gaspar Island, passing one or two miles East of that island; then along the West coast of Pulo

Leat, and enter the Java Sea through the Macclesfield Channel.

If at this period contrary winds are apprehended, a vessel before entering the strait of Gaspar may make the N.W. point of Billiton, round the islands off this point at a good distance, then steer West of the coast of Long Island, and thence, avoiding the dangers of the Clement Channel, would leave this channel between South and Selle Islands; she would afterwards pass East of the Vansittart Bank and the Dordelin Rock, and thence into the Java Sea.

At this season, on leaving Banca Strait, or that of Carimata, the strait of Sunda is the most direct route for the Indian Ocean.

Going up the China Sea.—In running up the China Sea from South to North with the S.W. monsoon, the outer passage should be preferred, East of the Paracels, where the sea is free from dangers. During this monsoon the wind draws southward, varying from S.S.W. to S.S.E. during the height of the monsoon (June and July.)

Although some *Directions* (amongst others those of Horsburgh) recommend the channel West of the Paracels and the coast of Cochin China, I think, says Commander Blake, that we should avoid it. He adds, that in this route a narrow channel must be crossed, and in some cases the risk is incurred of being carried by the current to the N.W. of the Paracels. Several vessels have been lost on the shoals in these parts, owing to the N.W. winds from Tonkin Gulf and the northerly currents. In the S.W. monsoon, a vessel taking the channel between Hainan and the Paracels, if the weather is cloudy, will find it very difficult, with currents varying in force and direction, to keep her course with sufficient precision to make a safe passage through it.

Horsburgh advises the inner route to be taken in case the vessel is disabled or has sprung a leak, because she can keep the land in sight.

Coasting Route up the China Sea.—On leaving Pulo Aor she would steer for the Redang Isles, then cross the gulf of Siam, passing Pulo Obi, and follow the coast of Cambodia or Cochin China, only leaving the latter near Tourane. From thence she will soon reach Hainan, and, running along the coast, leave the Taya Isles to the eastward, and then shape a course for the coast of China near Hai-Ling-Shan; from whence she would proceed along the coast to Macao. In this route a disabled vessel will find a smooth sea and pass many harbours, where, if necessary, she can take refuge. This passage would doubtless be tedious, especially about the middle of March or beginning of April.

Eastern Route up the China Sea.—The route for Canton through the middle of the China Sea becomes uncertain when Pulo Sapata has not been reached by the beginning of October. About the middle of this month strong southerly currents are found in the vicinity of that island. The light northerly and variable winds and calms which prevail at this period often prevent vessels from continuing their northerly route; in which case, having got clear of the shoals in order not to be longer delayed, a vessel should adopt the eastern route.

The southerly currents found near Pulo Sapata gradually decrease, and with the help of the favourable winds often found there, a ship may reach the parallel of 13° or 14° N. without much difficulty, where currents are not so formidable as in the vicinity of Cape Padaran, Pulo Sapata, or the Catwicks. A ship having attained the parallel of 13° or 14° will then be North of all the banks, and if the wind permit should get to the eastward, in order to make sure of crossing to Macao from the coast of Lucon. In this same season, also, the same route is adopted for Manila. The N.E. or E.N.E. winds, which often extend beyond the parallels of 12° and 13° N., render it difficult to make easting, and long boards should be made to the northward, and advantage taken of variable winds, to make easting, for it is not prudent to sight the coast of China West of the Great Ladrone.

Palawan Route.—A vessel late in the season should take the Palawan Route, because the southerly currents and the light breezes near Pulo Sapata and in the middle of the China Sea render that passage uncertain.

From the middle of November until February ordinary sailing ships generally make the voyage to China by the Eastern Route, or else enter the Sooloo Sea by the strait of Balabac, and, steering northward, successively stand along the coasts of Mindanao, Negros, Panay, Mindoro, and Luconia,—a route previously mentioned. According to Horsburgh, it is the shortest that can be taken after the beginning of November.

Steaming up the China Sea.—A steamer of small horse-power going from Singapore to Hong Kong during the height of the N.E. monsoon, should pass West of the Anambas; then, steering N.E., may pass the Natunas Islands at some distance, and then continue her N.E. course for the meridian of 115° E. She may then run to the northward and leave the Paracels to the West, and make direct for Hong Kong against the monsoon. By passing West of the Anambas, a smooth sea may be expected. Having reached the part before mentioned, going to the northward, strong breezes will be met, sometimes from the northward. Nevertheless, by sail and steam, and taking the best advantage of the wind according to its changes, a steamer will generally make the passage in ten days.

When the height of the monsoon is over, that is, in April, May, or June, a ship may take a more direct route than this. Leaving the Anambas to the westward, she might make for Pulo Sapata; then steer a little West of the Paracels; and then direct for China. If she meet with constant foul winds it will be useless for a steamer of small horse-power to attempt to steam.

There are others who consider the following a more advantageous route than those abovementioned. They say a ship should pass inside Pulo Aor, cross the gulf of Siam as far as the coast of Cochin China, and afterwards make to the West of Pulo Condore and Pulo Cecir de Mer; continuing then along the coast of the continent as far as the parallel of 15° N.; then making for Hainan, and keeping the land on board. In this season a strong current, however, sets southerly along the coast of Cochin China and the wind is no where light. Hence the former route, taken, during the height of the monsoon (N.E.), by the *Vixen*,—which vessel made the passage in ten days (9 days 23 hours) from Singapore to Hong Kong,—appears preferable. The distance is about 1,522 miles: the mean speed of this vessel would therefore be about 6·5 miles an hour.

Route for a Steamer from Hong Kong to Macao.—On leaving Hong Kong in a steamer for Macao, the channel between Hong Kong and Green Island should be taken, in which there is water sufficient for the largest vessel. She should steer for the North point of Chung Chou or Water Island, to pass between this island and Lantao. In the middle of this channel there is a large rock not shown on the chart; it is level with high water and is visible as it is approached. The vessel should pass between this rock and Chung Chou, and the least water will be six fathoms. She would afterwards take the channel between High Green and Lantao, and keeping in the middle of it, will not keep less than seven fathoms. Beyond here the chart will direct her, and this is the shortest route from Hong Kong to Macao.

A ship leaving Macao bound to the southward against the S.W. monsoon, should work down for the Macclesfield Bank, keeping between the meridians of 119½° and 116½° E., and profiting by every change of wind in her tacks. Having passed the Paracels, she would continue to the southward without making long boards to the westward, to approach the coast of Cochin China, on account of the strong northerly currents off that coast, which sometimes set fifty miles in twenty-four hours. She would therefore make small tacks, keeping the channel between the coast of Cochin China and the shoals East of it, taking care not to sight the coast of Cochin China at all, and not to stand further East than the meridian of the Investigator Shoal.

After passing Pulo Sapata, she would make small tacks, avoiding the entrance of the gulf of Siam, where light variable winds from the southward will be found. Sometimes, however, W.N.W. winds are to be had, and even N.W., but they are never of long duration. The Natunas may also be sighted to the southward, and then the channel may be taken between this island and the Anambas. It is better in general to take this channel than to sight the Malacca coast and the island of Timoan, off which light southerly winds with calms are found, which oblige vessels to anchor, so as not to be set to the northward by the current. A vessel will thus reach Singapore or the entrance of the straits leading to the sea of Java.

This passage may be made in thirty-three days, and the distance from Macao to Singapore is considered about 1,350 miles.

From the above it is evident that a good sailing vessel may work down the China Sea the whole way against the S.W. monsoon. At some periods, however, it is a passage not without great difficulties. In June, July, and August, for instance, a vessel of ordinary speed should not attempt it unless she can start with favourable winds on leaving the China coast; and even then, with a good vessel, bound to some western part of India, she should not take this route during those months unless compelled to do so. A vessel leaving China early may make a tolerable passage to Bengal by this route.

Nevertheless, if bound to Bombay a vessel will have a long passage from Acheen there in October and part of November. A vessel leaving China in May or June will probably reach Bombay as soon or even sooner by the eastern straits than by crossing the China Sea to take those of Malacca or Sunda.

From the Eastern Straits to Singapore.—The route for Singapore from Java, Bally, or Lombock, when the S.E. monsoon is far advanced, is often long and tedious on account of the S.E. current having commenced in October. After passing Pulo Mancap, a vessel often finds light breezes, frequently from the westward and N.W.

The following may prove useful to vessels taking this route, which is termed the North Route by Captain M'Kenzie in the *Nautical Magazine*.

After leaving Lombock and Bally Straits, the easterly winds will take a vessel beyond Pulo Mancap. The best route up to that point will be to pass between Pondy and Galion Islands: this may be done in perfect safety even during the night. The ship should then pass South of Lubeck Island; then keeping well to the southward and westward of the banks off Pulo Mancap and giving a good berth to the Discovery Bank and other adjacent dangers of the western part of the channel. She would then make for the eastern of the Montaran Islands, and take the channel between this island and that nearest West ot it, for it is perfectly safe. She would then steer N.W. along the North shore of Billiton, but should not pass within the banks of Montaran, because at this season of the year the wind is seldom more westerly than S.W.; consequently, a vessel would be in a sufficient position in starting from the eastern of the Montaran Isles, in order to pass to the southward of Pulo Docan, Pulo Taya, and Pulo Sinkep, especially the latter, if possible, in order to take the Brahalla or Great Channel. If not, the vessel should take the strait of Dasee, South of Lingin.

After leaving either of these straits, the vessel may make for Singapore through Durian Strait. On the West coast of Sinkep and Lingin, and in Dryon Strait, regular tides will be found, besides favourable changes of wind from the Sumatra shore. The Sumatra breezes are sometimes preceded by a calm, and if the tide is contrary the ship should anchor and wait a tide. These Sumatra breezes should be prepared for with double reefs, for a ship should not have to shorten sail and thus prolong her voyage.

On the eastern coast of Lingin, the current sets from the China Sea strongly to the southward. Besides, the wind here is often N.W. and North, so that a vessel can make but little progress until she has entered Rhio Strait. Again, it is frequently the case that a ship can only get on by working to windward, and is obliged to continue on a bowline all the way to Singapore. In this case it is better to take Dasee Strait (between Linga and the island North of it), instead of passing East of Lingin, in order to enter the strait of Rhio or Singapore.

From Singapore to Java by the Carimata Channel.— When a ship has to work through the Carimata Channel against the S.E. monsoon, the best route on leaving Singapore will be to pass through Rhio Strait, and then make for the coast of Borneo and work by short tacks along this coast, anchoring when the tide is against her. On leaving Kumpal or Rendezvous Island she would make for the coast of Java, and if going East she would work near the coast. The passage to Sourabaya can be easily made thus in fifteen to twenty days.

Vessels from the eastward of Java, Bally, or Lombock, bound to Singapore in the N.W. monsoon, may adopt either the southern or northern route. The one passes South of Java by one of the eastern straits, from whence the Trade winds will soonest be found, which will take a ship far enough West to enable her to reach Cape Acheen and cross the strait of Malacca;—the other crosses the Java Sea and the strait of Carimata. The mean length of these passages is about sixty days, and the strong winds and heavy seas are very trying to ships. Captain M'Kenzie prefers the Eastern Route to either of these.

By this route a ship reaches the Molucca Channel, or even Gilolo Strait, crosses the Celebes Sea, and the Sooloo Archipelago, and thence, with North or N.E. winds, passes through Balabac Strait to the China Sea, which is then crossed to Singapore. This route, which at first appears very circuitous, is however that by which the shortest passages are made, at least according to that officer.

From Amboyna to Banda.—A ship leaving Amboyna for Banda during the height of the S.E. monsoon (July, August, and September), should stand to the southward, even to the Timor coast: near which at this season the current sets easterly, and where the wind is found varying E.S.E. and S.E., by which Banda may be reached on the starboard tack. This passage frequently takes only six days. By going North of Ceram it would be much longer.

During the N.W. monsoon the route is direct from Amboyna to Banda.

Amboyna to Ombay Strait.—A ship leaving Amboyna in the N.W. monsoon, for the westward by the southern route for the Indian Ocean, should persist in passing North of Timor, and round it to the West. In fact, should the ship fall to leeward of this island, and be obliged to round it to the East, she will have great trouble in getting to the westward between it and Australia in the Timor Sea; where, from November to April, a heavy sea prevails, with strong westerly and W.N.W. winds and cloudy weather.

In case a ship is obliged to round Timor on the East, she should work along the South coast of the island and run through Limao Strait, for by keeping North of the Sahul Banks and near the coast, she will find less sea and less wind than out at sea.

Eastern Route from Amboyna to Malacca or India.—But the southern route from Amboyna or any of the Moluccas for the strait of Malacca or India, should not be taken in the height of the N.W. monsoon. For a vessel starting before March, the eastern route indeed is preferable, especially if she be going to the strait of Malacca and gulf of Bengal.

Adopting the eastern route from Amboyna, a ship should endeavour to reach the island of Amblau, and then work up for the East coast of Bouron, so as to profit by the land breezes and slants which are found there. She would thus avoid the southerly and S.E. currents which frequently prevail at this period near Amboyna, Ceram, and Manipa. Having reached Point Lisatika, (West end of Bouron,) she should make for Oby Major, towards the Gilolo Strait; and, having passed through this strait, should attain as soon as possible the parallels of 5° or 6° N., without caring to make easting, for in these parallels the N.E. monsoon prevails. She should then pass North of the Meangis, and make the South point of Mindanao, and pass between Mindanao and the Serangani Islands, and then run for Basilan Strait, by which to enter the Sooloo Sea. If this strait be entered in the morning with a favourable breeze, a vessel will clear it by the evening; if not, anchorage may be had at Samboangan. On leaving the strait of Basilan, the ship should pass to the northward of the Sangboy Isles, and then run for Cagayan Sooloo. From this island, avoiding the dangers to the northward and westward, the channel formed by Bengury Island and the Mangsee isles should be taken; from whence the ship would enter the China Sea, and then take the strait of Malacca.

On leaving Amboyna in March for India, the southern route should be adopted; and as during this month the wind frequently becomes more westerly, it will be better to make for Isle Bouron, and afterwards to stand for the islands of St. Matthew, for the purpose of working to windward under them. Having reached Velthoens Island, the ship should make to the southward, and if the wind permit take Aloo Strait, as preferable to that of Panthar; and having passed it, should make as much westing as she can, and pass between Sandal-Wood and Savu Islands, or East of this latter island if there be any difficulty in rounding its western end.

A ship leaving Amboyna from April to August, should stand towards Gunung Api, taking care to give a good berth to the Lusapara Isles. Passing East of Gunung Api, the ship should make for the western end of Wetta, in order to pass between Ombay and Pulo Cambing, and then take the Ombay Passage between Ombay and Timor.

A ship having thus reached Rottie or Savu Isles, will find S.E. winds gradually freshen. She should endeavour to pass South of Savu, from whence she may reach India in twenty or thirty days.

Moluccas to Batavia.—A ship from the Moluccas to Batavia during these months, should steer so as to sight Isle Roma, taking care to avoid the bank extending eight miles from its N.W. point, and then pass North of Wetta. From thence she should make for Pulo Cambing, and keep along the North coasts of the islands between Ombay and Flores. If desired, she may pass between this last island and Isle Rusa Raji; she will continue along the North coasts of Sumbama and Lombock, and from the latter island may pass between Hog and Galion Islands, or North of Urk, between the Four Brothers and Kangelang. From thence she would stand along the coasts of Madura and Java.

Navigation of the Gulf of Siam.—A ship from the southward, in the S.E. monsoon, bound to Siam, after leaving Pulo Timoan should keep out to sea until she has passed Pulo Varela. If the wind be favourable, she may pass inside of it, and should afterwards steer so as to psss inside of Pulo Brala, as the wind often hangs to the westward; thence she may take the channel between the outer Great Redang Island and that inside of it. Thence she may stand for the N.E. side of Pulo Lantinga and Pulo Printian, and thence make for the Siam River.

Leaving the Siam Gulf.—Vessels bound southward generally leave Siam with the N.E. monsoon, and if they start for the strait of Malacca before the middle of September, may expect long passages. In this case, a ship should work along the coast as far as the reef off Point Romania, so as to profit by the land winds, and in all cases be ready to anchor in a convenient depth if the current obliges her to do so, which is often the case in October with strong easterly winds, when the current sets up the gulf along the western coast. At all other times the rivers produce a current which takes an easterly or S.E. direction, and vessels in the S.W. monsoon, at the entrance of the gulf, are frequently set twenty or thirty miles to the eastward.

On leaving the harbour of Siam, in the N.E. monsoon, a ship should steer so as to be well clear of Cape Liant and its adjacent islands. Thence she should steer for Pulo Way, and not seeing these islands, should sight Pulo Panjang, and even although she may have sighted Pulo Way, it will be more prudent also to sight Pulo Panjang, from whence she would steer for Pulo Aor, if bound for the strait of Malacca or Banca.

On leaving the gulf of Siam with the S.W. monsoon, a vessel should work along the Malay coast if bound to the strait of Malacca or Banca. If bound to China, Cochin China, or Manila, she would follow the West coast of the gulf as far as Cin Point; from thence she would steer S.E. to sight Pulo Panjang, and if not made, should tack to the eastward until it is sighted, it being almost indispensable to do so. In this route she would also be careful to guard herself from the effects of the easterly currents which are met with in this season.

After Pulo Panjang is passed, a S.E. course should be steered for Pulo Oby, seventeen leagues, giving it a good berth in order to avoid the dangers near it.

Pulo Oby to Manila.—A vessel proceeding from Pulo Oby to Manila, should pass South of Pulo Condore, and thence South of Pulo Sapata; then N.E. until she has reached the parallel of 12° N.; and being then North of all the banks, may make for Manila. To pass the Redang Isles to the coast of Cambodia during the month of March, a ship should get as far East as she can, because at this time of year near Pulo Oby westerly currents prevail. At this season, also, the wind is from N.E. and E.N.E.

Macao to Manila.—Vessels trading between Macao and Manila continue to do so during both monsoons. On leaving Macao with that from the S.W. for Manila, every possible advantage is taken of the wind veering to S.E. or East, when they get to S.S.W. or South towards the Macclesfield Bank. From thence they are sure of reaching Manila; indeed, unless the wind becomes South or S.S.E., a vessel would reach Manila on one tack, and with the wind from South she would sight Goat Island, and the land South of the bay.

During the N.E. monsoon, when bound to Manila, they pass through the Lema Channel, and keep as far East as possible, making for the N.W. part of Luconia, towards Cape Bolinao. During this season every precaution must be taken against the current setting to leeward. Having reached the latitude of the cape or the coast near it, a ship should keep off it to avoid shoal water, and having passed the Sisters, should continue about four or six leagues from land till she is to the southward of the islands and the dangers off Point Capones. From thence she would continue along the coast for Manila Bay.

Manila to Macao.—The S.W. monsoon being fair for going from Manila to Macao, a vessel from thence would steer direct for the Great Ladrone, and with the wind at S.W. on approaching the coast of China she may keep this island N.b.E. or North. If the wind inclines East she should steer for the Great Lema and take that channel.

A vessel leaving Manila Bay for Macao, during the N.E. monsoon, should keep the land as far as Cape Bolinao, and from thence should stand out to sea if the wind permits her to make northing. But with N.E. and northerly winds, especially if the vessel is not a good sailer, she should keep along the coast as far as Cape Bojador before making for the coast of China.

Pulo Sapata to Manila.—During the S.W. monsoon, on leaving Pulo Sapata for Manila, a vessel should make the parallel of 13° N. by passing outside the shoals in the eastern part of the China Sea. Having reached the parallel of 12½° or 13°, she would make direct for Goat or Luban Island, which are both about twelve or thirteen leagues S.W. of the bay of Manila. Having passed three or four leagues North of Goat Island, she would steer for Mariveles Island or Corregidor, at the entrance of Manila Bay, keeping nearer the North than the South shore, and when six miles West of Corregidor, she would steer for that island, and pass between it and Tas de Foin. From thence it is only eleven leagues to Manila, and eight or nine to Cavite.

Manila to Pulo Aor.—The N.E. monsoon gives a fair wind for a vessel from Manila to Pulo Aor and the straits of Malacca and Banca. On leaving the bay she would steer West as far as the parallel of 12° N., and continue this course as long as she is not sure that her longitude is less than 111° E. When she is certain of being sufficiently to the West, she should steer for Pulo Sapata, allowing for the westerly currents, which at this season are sometimes very strong and set a vessel much out of her reckoning. Should Pulo Sapata not be seen on reaching its parallel, soundings should be taken, and then a course steered for Pulo Aor.

Route for the Gulf of Tonquin.—To reach the gulf of Tonquin or the river of that name from where she now is, she would take advantage of the S.W. monsoon, and would keep along the coast of Cochin China as far as Cape Chonmay, the South point of the gulf, or as far as Tiger Island. Thence she would steer North, passing the West coast of Hainan Island, at a suitable distance according to the soundings, which rapidly decrease as its surrounding banks are approached. These should not be found in less than 14 or 17 fathoms. A ship not making Hainan Island when in lat. 19° N. and in 24 or 26 fathoms water, should steer for the Norway Isles, about thirteen leagues from the entrance of the Tonquin River. In taking this route, when approaching the entrance of the river, she would steer according to the tides, which may greatly retard her course, setting her out of the gulf.

Gulf of Tonquin.—During the N.E. monsoon, on leaving the Tonquin River, a ship should steer so as to sight Hainan Island, but should not approach this island from the N.W., and should take care not to get into less water than 17 fathoms on the banks, which extend from eighteen to twenty-one miles from its shores. Easterly winds often prevail in the gulf during the N.E. monsoon, although off the mouth of the gulf at the same time the true monsoon may be found.

CHAPTER XIV.

THE COAST OF CHINA NORTHWARD OF THE CANTON RIVER,—ROUTES TO BE TAKEN.

Passages along the Coast of China from South to North.—In making to the northward against the N.E. monsoon, along the East coast of China, from Macao to the Chusan Isles, several routes may be adopted. The first is by working to windward, between Formosa and the coast. The second, passing East of this island and working along its outer shore; and a third, by keeping well out to sea, eastward of Formosa. This last route Commander Goldsmith says is the best, and he tells us that in eighteen or twenty days a vessel may make the passage by it from Hong Kong to Chusan. This passage against the N.E. monsoon generally produces much injury to sails and rigging, on account of the constant succession of bad weather. There is generally a breeze strong enough to oblige a ship to double reef, while at the same time she has a heavy sea to encounter.

A ship bound northward having left Hong Kong during this monsoon, if the wind permits, may leave by the Ly-ee-Moon Channel, and pass the Nine-Pin Rock, and work along the coast, under Single Island and the group to the northward of it. Here the tide, during the syzygies, flows N.E. between one and two miles an hour. The wind is generally from E.N.E., or else comes down along the coast, veering more northerly when taking its direction, and remains steady at N.N.E. as soon as the Lamock Islands are passed.

Having passed Single Island and the Acoong Chow, a remarkable sharp pointed rock, she may stand into Harlem Bay, and pass Mendoza Island. Care should, however, be taken not to approach its South coast at less than three miles from Ty-Sing-Cham or Pedro Blanco, in order to avoid a dangerous rock, which only shows at low water by a heavy break.

The ship, if still going North, should keep near the China coast as far as Ma Ush, in order to profit by the tides and avoid the current, which is constantly setting to the S.W., and when the weather is uncertain, it is even better to anchor under shelter of one of the outer points of the island than to stand out to sea.

In the bay of Ti-Sami, a little West of Hie-Che-Chin, and under the point of Capehi, there is a good anchorage in 8 or 10 fathoms. At night care should be taken to give the edge of the breakers a berth of two miles, for banks and rocks extend from under water, and it must not be approached.

Having reached Cape Ma Ush, the route to be taken must be decided on, whether it be to continue North along the China coast, or to proceed East of Formosa. Sometimes the coast of China is left before reaching Cape Ma Ush by a ship taking the outer route East of Formosa. But is not difficult at any season to round this island to the East, because on the outside of it there is a current flowing North at a rate of two and a half or three miles per hour.

First Route—through the Strait of Formosa.—By the route along the coast of China, a vessel will find some difficulty in getting to the northward from the Lamock Isles. As far as those called the White Dogs constant southerly currents are encountered, which in the middle of the strait, and according to the strength of the monsoon, attain a rate of as much as three miles an hour. A ship would therefore adopt short tacks along the eastern part of the Lamocks, and then keep as near as possible to the land from the point of Jokako to the environs of Tongsang Bay, the shores of which are the best along the coast, and may be approached without danger. The little group of islands situated North of Rees Channel, forms a shelter from the sea that she may take advantage of; and if at night the ship should be to the East of these islands, it will be wiser to anchor under their lee than to run out to sea in this latitude, where the tides are strong and the direction of the currents very uncertain. In the dark nights, too, when North of these islands, a ship should not stand at all into the bay of Hoo-Tow-Shau, as several vessels have been lost on the rocks off its entrance. We may consider this part, says Captain Mundy, as the only dangerous one in the whole of this route. The rocks, called the Two Brothers, are a good mark to recognize, being detached from the coast, and without dangers near them. From thence the ship may keep along the coast towards Red Bay, but there are several single rocks to be avoided, although no dangers have yet been discovered under water, if we except a shoal situated three quarters of a mile from the coast, and three miles South of the anchorage of Red Bay. From this anchorage, in fine weather, Chapel Island will be seen, which in the day time may be passed either to the East or West, taking the necessary precautions to avoid the Merope Shoal.

After having rounded Chapel Island, the islands at the entrance of the port of Amoy will be seen. If the weather is cloudy and the wind fresh from the North, the North coast should be kept on board, and with a smooth sea, which will be always the case with the wind from North or N.E., anchorage may be taken in 30 to 40 fathoms. Continuing to the northward as near the land as safety will allow, the ship would enter the bay of Lee-o-Loo, passing near Dodd Island, and then into the bay of How-e-Tow, anchoring during the night, or, if obliged to do so, by the tide.

The pagodas of Chinmo and Chin-Chew, both visible at some distance, are good objects to distinguish in this part for rectifying the position. On losing sight of the latter, when Cape Tong-Boo is passed, less sea will be found and the ship may stretch out further from the land up to the isles of Ock-Seu, three barren rocks, under the lee of which there is no shelter except close to them. Having rounded them, the island of Lam-Yit may be hugged close. In these parts, during the whole of the N.E. monsoon, heavy winds are encountered, with squally weather and a strong current to about ten miles from land, so that in order to get to the northward a ship must again make short tacks for the Three Chimneys, taking care not to go too close to the reef islands or to Cape Double. She would pass West of Isle Turn-about, a remarkably good landmark.

To pass this point is perhaps the greatest difficulty of the whole route, and if the monsoon came with but half its strength, the ship might be driven away South of Ock-Seu, unless she anchored under the lee of one of the islands while the lee tide is running, or else kept entirely under the land. The swell also being heavy, the ship is strained and twisted about by carrying a press of sail.

When she has rounded the Three Chimneys, the first islands seen are the White Dogs, which may be passed on either tack, according to the weather, for Matson Island. But by making short tacks the vessel will keep a smooth sea until passing Chang-Chi. Off the South part of this island there is a good anchorage, which should be made use of by a vessel intending to communicate with Fou-Chow-Fou, but which she would do sooner by boat, leaving the vessel anchored at the White Dogs. If the weather is tolerable, there will be no more swell after passing Chang-Chi, and when the Trio Rocks are passed, the ship would work between the islands of Larne, Alligator, and Tong-Yang, taking great care to avoid the Larne Rock, although it is always visible. Between these islands there is anchorage in 17 or 18 fathoms. In the syzygies the current of tide attains a rate of three miles; but at neaps the tide is not more than one and a half mile per hour. The ebb sets from West, then changes to S.W. and South. In the morning the wind generally veers northerly.

From the last mentioned group the island of Pih-

M

Seang should be made, and from this as far as Kew-Shan the ship may work up the coast, leaving the different groups of islands inshore of her to the West, until she reaches Hih-Shan. Either by day or night she may stand without fear between these islands and the coast; but when she has reached Patahecock, (elevation 450 feet,) the most southerly of the Kew-Shan Group, she would not pass West of it excepting in the day, as there are dangers in the channel, and the approaches to Cape Montague are not to be trusted.

Between this cape and Patahecock there is a very dangerous rock, which the ship should pass East of, and then work between the sunken Pylades Rock and Bear Island. Some rocks, called the Whelps, would then be seen, and then the Corkers, a very remarkable group of rocks. They might be rounded on either side and on either tack. If the ship works East of Kew-Shan, she would pass near the Mouse, a low rock, covered at high water.

In this part the tide sets N.W. and S.E.; and good anchorages may be found from 4 to 8 fathoms.

These directions are given in detail because they will afford some idea of the navigation of the China coast, for it must be remembered that a good passage can only be made by keeping the coast on board.

Second and Third Routes,—East of Formosa.—A ship adopting the second route, East of Formosa, on leaving the Lema Isles should keep the coast, as above stated, taking advantage of every favourable circumstance to make easting. At the same time she would avoid the adverse current, which she would find out at sea, by keeping by day in the tranquil waters of those bays, which are safe; and if the wind is too strong, it would be wise, when near a good anchorage, to keep it, until it becomes less. In fact, she would make to the eastward when the wind admits, and get shelter from the land when it is foul. This is, in reality, the most difficult part of the passage, and requires constant attention.

The ship having reached far enough along the coast, that is, when, according to some commanders, she has reached Ma Ush, she would make for the S.E. point of Formosa. With the wind at N.E. they would leave the coast of China before reaching this cape; but however this may be, having doubled it the ship may then, according to the wind, pass either East or West of the Meia-co-sima Group.

This second route has the advantage of a favourable current, setting strongly to the northward of Formosa. When the wind at sea hauls to the eastward of North, which is generally followed again by strong winds from N.W., the ship would adopt the most favourable tack. But she should by all means avoid the land until she is North of Kew-Shaw, and then if desired may find a good anchorage under the shelter of these islands.

Mr. Foster, in alluding to these routes, says, that the currents of the strait of Formosa should deter any ship from taking the first route, and that the route outside Formosa, or to the East, is generally more advantageous during the N.E. monsoon. He also adds, that even leaving Amoy for the Chusan Isles a vessel should pass East of Formosa unless she is unable to work out of the strait. It is an opinion that should be followed.

CHAPTER XV.

FROM INDIA AND CHINA TO AUSTRALIA,—AND RETURN.

From India to the East Coast of Australia.—North Route, through Torres Strait.—This might be adopted from the beginning of September to the end of March, the period of the N.W. monsoon in those latitudes: the other the South Route, by which a ship passes to the South of Australia and through Bass Strait, is adopted from March to September, when the S.E. monsoon prevails in Torres Strait.

The navigation of Torres Strait belongs to that of the Pacific Ocean, and is of a local character.

For the southern route it will be easy to deduce directions from those we have previously given. The ship from the northern part of the Indian Ocean, should endeavour to gain as speedily as possible the zone of the Trade winds, and should cross this zone as soon as she can, obliquely inclining to the West coast of Australia. She would then keep along the South coast for Bass Strait.

If coming from the Cape of Good Hope she would take the same route as that described as the great eastern passage to China.

From China to the East Coast of Australia.—A ship leaving Canton during the S.W. monsoon for the East coast of Australia, the shortest route generally is the eastern one by entering the Pacific Ocean South of Formosa; she would then make sufficient easting to enable her to stand across the Trade winds from N.E. and S.E., so as to find herself to the East of the port of Australia to which she is bound by the time she has gained its latitude. This route presents no other difficulties than those of sailing among the islands of the Pacific Ocean.

This will be, however, very tedious and difficult work in the N.E. monsoon; for the ship must take the Pacific by the strait South of Formosa, and then work to the eastward and northward against the N.E. Trades, and look for the variables to the North of the parallel of 30° N., with which she may get to the eastward, therefore on leaving the China Sea she should get out of the Trade winds as fast as possible. As a rule she would stand as far as the meridian of 165° E. before running South.

But during the N.E. monsoon, that is, from September to February, a ship would find it more advantageous to keep in the China Sea, and would avoid a great deal of bad weather, besides making a quicker passage. She should pass East of the Great Natuna, and then take the Carimata Channel, and would next make for the East point of Madura, and pass between Pondy and Galion Islands, if deciding to take Bally Strait. It is, however, preferable to take that of Lombock, and more so still to adopt Allass Strait.

A ship leaving these straits should stand to the southward, profiting by the changes of wind, which in this season are generally S.S.W. and S.S.E. in the space between the limit of the Trade winds and the eastern straits. The ship should therefore make a long stretch even westerly to cross the zone of the Trade winds, and get into the variables of the southern hemisphere, and when she has reached them she will soon make up her easting. In March and April she may pass through Bass Strait, or, if preferred, South of Tasmania, as easterly winds prevail in that strait during these two months. At all other times Bass Strait would be adopted as the most direct and shortest route.

Ports on the East of Australia to India,—Southern Route.—A ship leaving Port Jackson or any port on the East coast of Australia for India, may take the route through Bass Strait or round the South coast of Tasmania, when commencing her voyage between the beginning of September and the first of March. In January, February, and March, near Tasmania, she will find S.E. winds more prevalent than during all the other months. She will profit by these winds in making to the westward, but must give a wide berth to the Australian coast, in order not to lose by a change of wind, and not to be caught too near the shore by heavy S.W. squalls, which prevail at this period.

Some vessels bound to Europe also take this route during the winter months, June, July, and August. It is true they suffer much from bad weather, but the passage is not impossible even during this season. As much as possible she would endeavour to reach the Trade winds, and when found would shape the most convenient course for the port to which she is bound. Under favourable circumstances a passage from Port Jackson to Bengal may be effected in two months; some vessels have been fifty days in going from Bass Strait to False Bay; but during the summer months (October, November, and December) the best passages have been made.

Northern Routes, through Torres Straits.—From March to September the passage from Port Jackson to Bengal or ports of the inland seas, is made through Torres Strait. In this route the ship runs to the northward along the coast of Australia as far as Sandy Cape; and then takes the inner passage. A ship taking the outer passage on leaving Port Jackson would make the best of her way to the meridian of 155° and should then pass West of How and Middleton Islands; then stand on to the North, keeping a look out for Wreck Reef and the Bampton Shoals, the Minerva, Bellona, Ball, and Kenn Reefs, &c.; which must be passed between Lamb and Bass Islands, all lying between the parallels of 24° and 20° S. Great care must also be taken to allow for a N.W. or westerly current, of at least one mile an hour, sometimes more. When Wreck Reef is passed, the course should be shaped so as to pass about sixty miles East of the Diana Bank for Torres Strait.

Port Jackson to India, East of New Guinea.—The best route for a ship bound to India or China from Port Jackson, not taking the southern route by Bass Strait, nor the northern by Torres Strait, will be to pass East of New Guinea and take Pitt Passage, and then cross the Java Sea or take the strait of Ombay. But this route, like the northern route, should only be taken from March to September, that is, when the S.E. monsoon prevails South of the equator. On leaving Port Jackson or Tasmania she would make to the eastward, keeping also a little North till she reaches the meridian of 160° E.; then stand to the northward, keeping nearly on this meridian, and passing away East of Howe and Middleton Isles, Wreck Reef, and the Cato Shoals.

Captain Bristow considers it preferable, on leaving Port Jackson, to stand to the northward as far as the latitude of Sandy Cape, passing West of the abovementioned dangers. A ship would thus pass New Caledonia on the East, taking great care to avoid the dangers of those latitudes, and then steer for Cape St. George, the South point of New Ireland.

If desiring to enter the St. George Channel and Pitt Passage, the ship should keep well in to the coast of New Guinea, to avoid the easterly currents and light N.W. winds that are frequently found North of the equator, and which set vessels out to sea. Having therefore reached the meridian of 134°, she would keep near the coast of New Guinea when approaching Pitt Channel in the S.E. monsoon from March to September.

Instead of taking the St. George Channel the ship may use that between New Britain and King William Cape, passing on either side of Rooke Island; from hence she would steer so as to pass between Lottin and Loupee Islands, or this last may be rounded on the West, according to circumstances. She would then continue along the coast of New Guinea, passing North of the off-lying islands from Astrolabe Bay as far as Isle

Dumont Durville, and continue along the coast as far as the point of this name. From this point she would stand on so as to pass North of Traitor Islands, Mysory, and Providence, and then make the coast of New Guinea at Cape Mamori. steering along the coast as far as Cape Good Hope, and then make for Pitt Channel through Dampier Strait. This route is shorter and more direct than that by St. George Channel, and there are fewer dangers to guard against in it.

The ship having reached Pitt Channel through Dampier Strait, or by any that appears more suitable, would be guided by the general directions previously given for proceeding to China through Pitt Channel, passing East of the Philippine Islands.

When the N.W. monsoon prevails South of the equator, a ship bound for India should not keep near the coast of New Guinea as above directed. In November, December, and January, it is better to get to the northward as far as latitude 5° N., near which the N.E. Trade may be expected. At this season, also, the ship would pass South of Mindanao, through Basilan Strait, and cross the Sooloo Sea, and then enter the China Sea by the strait of Balabac, North of the isles of Benguey and Balambangam. Then crossing the China Sea she would take the strait of Malacca. This route is doubtless the best for India during the N.E. monsoon from the East coast of Australia or Tasmania.

Eastern Routes for China or India, from the East Coast of Australia.—A ship from the same coasts bound to China or India, while the N.W. monsoon prevails South of the line, (from September to March,) and no adopting the seuthern route by Bass Strait, would encounter heavy winds if she takes the eastern route, that is, a route still more easterly than the last mentioned. In this case either one of two routes is generally adopted; the first passing West of New Caledonia, the New Hebrides, and the Santa Cruz Islands, and East of the Solomon Isles. The second passing East of New Caledonia, the New Hebrides, and Santa Cruz Islands. These two routes are nearly the same as those given for the Great Eastern passage from the Cape of Goed Hope to China.

First Eastern Route.—A ship adopting the first of these routes, on leaving Port Jackson would steer E.N.E. to profit by the regular breezes prevailing out at sea, and having reached the meridian of 160° E. would then pass to the eastward and northward of Howe and Middleton Isles; and as off New Caledonia S.W. winds are often found, it must not be approached too close. After rounding this island to the westward the ship would continue northward, keeping on the meridian of 164° so as to enter the channel between the islands of Santa Cruz and the Solomon Islands. This group being passed, if in the first months of the monsoon the ship is to be in China, a northerly course should be adopted to cross the Caroline Archipelago, having cleared which a course must then be shaped for Guam Island, reaching which one of the channels of the Marianne Islands would be taken for the Bashee Islands Channel.

In case the ship should not reach the Solomon Isles till after January, when the N.E. monsoon is less strong in the China Sea, she may pass between the islands of Guap and Goulon, or between this latter and the Pelew Isles. From thence she would round the N.E. point of Luconia, and take the most suitable channel for crossing the strait of Formosa and reaching China.

Second Eastern Route.—The second eastern route from the eastern coast of Australia or Tasmania to China is longer than the foregoing, but has not so many dangers to be avoided, besides possessing the advantage of more regular winds than those found West of New Caledonia and the New Hebrides. But in passing to windward of these islands the westerly currents found in crossing the zone of the S.E. Trade winds must be carefully provided for.

On leaving port it will be best to make Norfolk Island, or else to pass a short distance West of it, and then make for Isle Matthew, which may be seen at the distance of twenty-four miles. The ship would then pass East of New Caledonia, the New Hebrides, and the neighbouring isles, leaving those of Viti to the eastward, and giving the New Hebridres a good berth, would keep on the meridian of 171° or 172°. If the ship cannot round Erronan Island, the easternmost of the latter, the channel separating it from Tanna may be taken, from whence she would steer North as far as latitude 13° S. If the New Hebrides have not been sighted the ship must endeavour to mak Tipokia Isles or Fatika, and then continue her northern course to cross the equator between 160° and 168° E. As the current generally sets strong to the westward, the ship should endeavour to reach the Carolines on the meridian of 163°, and if she has crossed the line in 160° or 162° E., she should endeavour to cross this archipelago between the meridians of 154° and 153°. On leaving the Carolines, she would pass South of the Mariannes, or through one of the channels formed by these islands, and then make for the channel South of Formosa to enter the China Sea.

In the tracks abovementioned should any unforeseen circumstance prevent a ship from reaching China before the end of the N.E. monsoon, when she is North of the equator the latter part of the route should be changed. She should then pass South of Mindanao and take Basilan Strait; she may also in this case cross the Philippines, taking Panay or Surigao Strait, or that of St. Bernardino. The first is North of Mindanao, the second North of Samar. By the first the Sooloo Sea is entered; by the second the China Sea is entered North of Mindoro. She may also pass South of this last island, taking the strait of Mindoro. In entering the China Sea from the Pacific Ocean the strait of St. Bernardino is generally preferred to that of Surigao.

The Inner Passage to Torres Strait.—Captain King gives the following directions for this route:—

Having hauled round Breaksea Spit in the evening, it would perhaps be dangerous to steer on through the night. After running, therefore, to W.N.W. for five or six leagues bring to until daylight. But if the day be before you the course from the extreme of the spit is W.N.W. ¼ W. for about a hundred miles. You will then be about twenty miles from Cape Capricorn; on your way to which you should pass about three miles within Lady Elliot Island (1), and also within the southernmost islet of Bencher Group (2), by which you will see how the current has affected your course, and you can act accordingly. If it has set you to the northward you can pass on either side of or through the islands without danger. After making Cape Capricorn, you may leave it at a convenient distance and, directing your course at about N.W.b.N., pass either within or without the Peaked and Flat Islands off Port Bowen (4); then, steering for the Percy Group (6), pass between the second and third Northumberland Islands (5).

After passing the latter, avoid a low dangerous rock that bears from it N. 8° E. 5¾ miles, and from the first peak S. 85° W. To avoid this in the night, pass close round No. 3, which is high and of bold approach.

The channel is safe on either side of the Percy Isles, but that to the westward of them, being better known, is recommended as the safest. Following either the *Mermaid's* or *Bathurst's* track will carry a ship round the projections of the coast as far as Cape Grafton (35); as far as which, if the weather be fine, there can be no danger of proceeding through the night. But it must be recollected that at Cape Grafton the coral keys of the Great Barrier Reef approach the coast, and consequently the channel becomes contracted.

On reaching Fitzroy Island (36), round it a mile off, and when its North end bears West, steer N.W. ½ N. for thirty-five miles; you will then be a league to the S.E. of a group of low isles. Should it be night when you pass them, come no nearer to them than 14 fathoms. In steering this course great care should be taken not to go too much to the eastward, to avoid the reef which the *Tamar* saw.

If the moon be up and the weather fine, the islets will be readily distinguished, but it would be more prudent to wait for daylight. This course will carry a ship over two of my tracks, and the soundings will be 17, 18, and 19 fathoms.

From the low isles direct your course for the Hope Islands (37), which bear from the former N. 18° W. thirty-eight miles; but the course had better be within (West of) that, to avoid some reefs in lat. 15° 51′: pass, therefore, within five miles of Cape Tribulation, whence a direct course may be steered on either side of the Hope Isles. The better route will be within the western Hope and along its reef, at the distance of three quarters of a mile, by which you will avoid reef *a*. When you are abreast of its North end, a N.W.b.W. westerly course for twenty-eight miles will carry you to Cape Bedford (40), which may be rounded at from one to three or four miles. You will see in your way, at three and a half miles from the North end of Hope Reef, reef *b*; and at fifteen miles from it you will be abreast of *c*; and five miles further on you will pass Captain Cook's Turtle Reef (43), which has a dry sand at its North end. These three reefs will be to the eastward of your course.

The current sets to the N.W., so that your course must be directed accordingly. In coasting along shore, you will discern the summits which are marked on the chart. The high conical hill on the South side of the entrance of Endeavour River (38) is Mount Cook; bearings of which, crossed with the summit of Cape Bedford, or any of the particularized summits or points, will give the vessel's place; by which the effects of the current, which is generally very slight, will be ascertained. On one occasion we found a current in the space between the Endeavour Reef and Turtle Reef of two miles an hour to the N.W.

Being off Cape Bedford, and steering to the N. ½ W., you will see the three isles (44) ahead. Steer between them and the low wooded island, and direct your course round Cape Flattery (41) and Look-out Point (42), in order to anchor under the Turtle Group (49), unless you have time before dark to reach the islands 4, 5, or 6 of Howick Group (52), under which anchorage may be found. In rounding Look-out Point, do not come within two and a half miles of it, to avoid a reef that is on Captain Cook's chart, which, however, we did not see; it lies a mile and a half North from the peaked hill at the extreme of the point. You may pass on either side of the Turtle Group, or between Lizard Island (46) and Eagle Island (48), but the latter course is not to be recommended: first, because the wind is generally fresher as you increase your distance from the shore; and, secondly, because the run is ten or twelve miles longer. There is good anchorage under the N.W. side of the peak on Lizard Island.

From the Turtle Group steer N.W.b.W. ½ W. until you see the hillock at the S.E. end of Howick Group (No. 1); pass inside it and within a mile of 2 and 3, and between islet 4 and Coles Islands (50), also inshore of 6 and of the dry sands *s*, *t*, *u*. The *Mermaid's* track will direct the course to Cape Melville (56). If the day be advanced when abreast of 6 of Howick Group, anchorage had better be secured under it, as there is none to be recommended between it and Cape Flinders (59).

Upon rounding Cape Melville, the islands of Flinders Group (58) will be seen; and as soon as you have passed between the stoney reef that projects off the cape (the extreme of which bears from it, by compass, N.W.b.N. and from Pipon Island S.W.b.W. ¼ W. nearly) and the reef that surrounds Pipon Island (57), direct the course for the extreme of the islands to the westward, which is Cape Flinders, the course and distance to it W. ¾ S. nearly thirteen miles, leaving a low woody island on the starboard hand.

H.M. sloop *Satellite*, in 1822, grounded on a small reef, *a*, bearing N.b.E. (easterly) from the extreme of the cape. This danger may be easily avoided as a ship may pass within a stone's throw of the cape. The best anchorage to leeward of Cape Flinders is under the flat topped hill, at one third of a mile from the shore, in ten fathoms, muddy bottom. In hauling round the cape, avoid a shoal which extends two cables from the shore on its western side.

If daylight will allow to run fifteen miles further, the ship may proceed to the anchorage under reef *d;* but in this neighbourhood anchorage may be obtained under any of the reefs or islets between this part and Cape Grenville, for the bottom is everywhere mud, and by anchoring with the body of a reef bearing S.E., the vessel is sheltered from the sea, which is generally smooth.

On leaving Cape Flinders steer W. ¾ N. for about twenty-three miles, leaving the reefs *c* and *g* to seaward, and *d*, *e*, *f* to the southward of the course. Then haul up N.W. $\frac{3}{4}$ N., and steer within the reef 1 and Pelican Island (63), and to seaward of the Claremont Isles (64) 1 and 2, which are low and woody.

When abreast of island 2, the S.W. end of the reef *m* will be seen, which should be passed at one to two miles, and the course N.b.W. $\frac{1}{4}$ W. will carry you to islands 4 and 5, which you may pass on either side,—the channel between them is also quite safe. If you take the latter course, steer North, within the reef *o*, to avoid a low rock that covers with the tide. Having passed this rock, which is marked on the chart, steer for 7, and pass within one mile of it, to avoid the shoals that extend off Cape Sidmouth (66). Hence the course is N.N.W. towards Night Island (67), and when abreast of it steer N. $\frac{1}{2}$ W. until near the covered shoal *v*, when the course may be steered within Sherrard Islets (68) and reef 10 (on which there is a sandy islet covered with some bushes), and then steer round Cape Direction.

Hence a course N.N.W. $\frac{1}{4}$ W., will carry you within the reefs *y*, *z*, *a*, *b*, and *c*, and without the rocky islet that lies off Restoration Island (70). Continuing this course for about five miles beyond Cape Restoration you will see the long reef *e;* then steer N.W. along its edge, which extends until you are abreast of Fair Cape (71), where it terminates with a very narrow point. Thence steer N.W. $\frac{1}{2}$ N. and pass between the two easternmost Piper Islands (74) and the reefs *h*, *i*, and *k;* then on either side of *l* and *m*, inshore of Haggerston Island (76), and round the outermost of Sir Everard Home Group (77).

The anchorages between Cape Flinders and this are so numerous as not to require particular mention. The

N.W. end of every reef will afford shelter, but the anchor should not be dropped too near them, because the tide sweeps round the edge with greater strength than it does at half a mile off, and the bottom is generally deeper. If the day is advanced and the breeze fresh, Night Island should not be passed, because the anchorages between it and Piper Islands are rather exposed, and a vessel getting under way from Night Island at daylight will easily reach Piper Islands or Margaret Bay before dark.

Margaret Bay (77) is round Cape Grenville. The anchorage being fronted by Sunday Island, is well sheltered; it is a safe place to stop at; anchor in six fathoms, with the island bearing S.E.

In passing round Sir Everard Home Islands (77) steer wide from them to avoid the tide drifting you towards the group, for it sets to N.W. across the course. The course is then about N.W. ¼ W. to the Bird Islands (80), and then to the reef *v*, about N.W.b.N. The better and more direct plan is to pass within *v* and *w* (there is, however, a safe channel between them), and when abreast of the West end of the latter, the course and distance to Cairncross Island (82) is N.b.W. ½ W. about eighteen miles.

There not being any very good anchorage between Cairncross Island and Cape York, it would be better perhaps to anchor under it for the night, in about fourteen or fifteen fathoms, mud, the island bearing S.E., but not nearer than half a mile, because within that distance the bottom is rocky.

Leaving Cairncross Island steer N.N.W. ¼ W. until

Escape River (84) is abreast of you, when look out for reef *x*; steer within it about N.W.b.N., which will take you inside the covered reef *z*. Your course must then be round the Albany Islands (87), and hence N.W.b.N. for *a*, which is a rocky islet that may be seen from abreast the Albany Isles.

The passage through the Possession Isles (90) and Endeavour Strait (91) is not to be recommended for a large ship on account of the shoal water that extends from Wallis Isles towards Shoal Cape: the route round the North end of Wednesday (92) and Hammond Islands is preferable. On passing reef *a* Wednesday Island will be seen. In steering towards it, avoid standing too close to the rocky islet that is abreast of the strait between it and Horned Hill, as some sunken rocks stretch off it for about a quarter of a mile; steer round the North point of Wednesday Island at half a mile, then W.b.S. ¼ S., which will carry you to the northward of the rock off Hammond Island. From this rock steer S.W.b.W. and when abreast of the S.W. end of Hammond Island haul towards a reef, to the southward of the course, on which you will see some dry rocks, which you may pass within half a mile. By so doing you will avoid reef *d*, which is generally if not always covered. The fairway of this channel is seven and eight fathoms deep.

When the summit of Good Island (92) bears S.W.b.W. steer W.b.S. southerly for Booby Island (93), by which you will avoid Larpent Bank, and when you are past it you are clear of the strait. Hence you may steer W. ¾ S. through the night; on which course you will very gradually deepen your water.

Captain King makes the following further remarks:—

The season in which the strait can only be passed commences with the month of April and ends with October,—that is, during the S.E. monsoon. The westerly monsoon, besides being a foul wind, is accompanied by gloomy unsettled weather and heavy rains, and frequently by strong gales, against which it is impossible to make a voyage without great delay and loss of time.

The passage up the N.E. coast is not affected by the monsoons. The S.E. Trade blows up the coast during the whole year with little variation, save that during the months of June, July, and August it is occasionally, although rarely, suspended by N.E. winds, with thick, rainy weather. To the southward of Cape Grafton (lat. 16° 51′) this interruption is of more frequent occurrence. Still, however, the prevailing wind to the northward of Breaksea Spit is from S.E., and there will be no difficulty in making an expeditious passage.

PASSAGES PERFORMED BETWEEN ENGLAND AND AUSTRALIA.

We shall conclude with the following summary of passages out and home between England and Australia, and the average duration of voyages in the years 1856 and 1857. The shortest have been made by ships of the largest tonnage.

To Victoria, South Australia, the shortest voyage was made in 68 days by two ships from London of above 1,000 tons, and the longest was one of 144 days by a ship of 426 tons.

To Sydney, the quickest passage was made in 74 days by a London ship of 1,500 tons, and the longest by a vessel in 168 days, although an exceptional case.

The runs homeward are from Melbourne and Sydney 70 days.

The returns from which the passages are given tend to show that ships of large tonnage, as might be expected, possess, with but few exceptions, great advantages over smaller vessels, although much depends on the skill and ability of the commander. In nearly every case London ships have proved the swiftest, but Liverpool can boast of despatching some of the finest,—one of them, registering 2,143 tons, running out to Sydney in 75 days.

London to Victoria.

Years.	Ships.	Tonnage.	Average Tonnage.	Av. Pass. Days.	Longest. Days.	Shortest. Days.
1856 ...	108 ...	89,197 ...	826 ...	102¼ ...	129 ...	71
1857 ...	127 ...	110,663 ...	871¼ ...	102½ ...	144 ...	68

Liverpool to Victoria.

1856 ...	76 ...	93,492 ...	1,236 ...	94 ...	130 ...	69
1857 ..	99 ...	125,885 ...	1,272 ...	94¼ ...	142 ...	71

Victoria to London.

1856 ...	24 ...	17,666 ...	736 ...	92½ ...	123 ...	81
1857 ...	30 ...	23,903 ...	797 ...	97½ ...	124 ...	70

Victoria to Liverpool.

1856 ...	20 ...	26,535 ...	1,327 ...	90¼ ...	128 ...	75
1857 ...	18 ...	32,278 ...	1,793 ...	100¼ ...	138 ...	76

London to Sydney.

1856 ...	69 ...	59,871 ...	867 ...	103½ ...	144 ...	74
1857 ...	73 ...	62,949 ...	862 ...	105¼ ...	145 ...	83

Liverpool to Sydney.

1856 ...	20 ...	19,555 ...	977½ ...	109¾ ...	168 ...	75
1857 ...	28 ...	23,129 ...	822½ ...	113¾ ...	153 ...	86

Sydney to London.

1856 ...	38 ...	29,155 ...	767 ...	107 ...	146 ...	79
1857 ..	29 ...	26,664 ...	919 ...	98¼ ...	126 ...	70

LONDON:
WALTER SPIERS, PRINTER, GREAT PRESCOT STREET.

CHART OF THE PREVAILING WINDS AND CURRENTS IN THE INDIAN OCEAN.

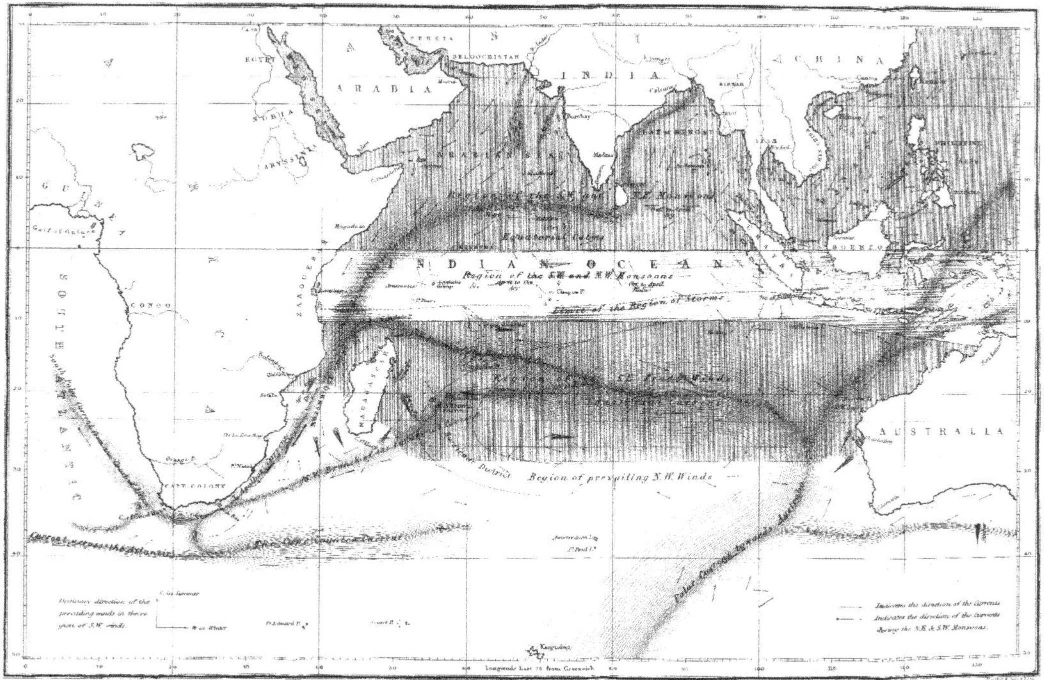

Sold by J. D. Potter Agent for the Admiralty Charts 31 Poultry & 11 King St Tower Hill

TRACK CHART OF THE INDIAN OCEAN.

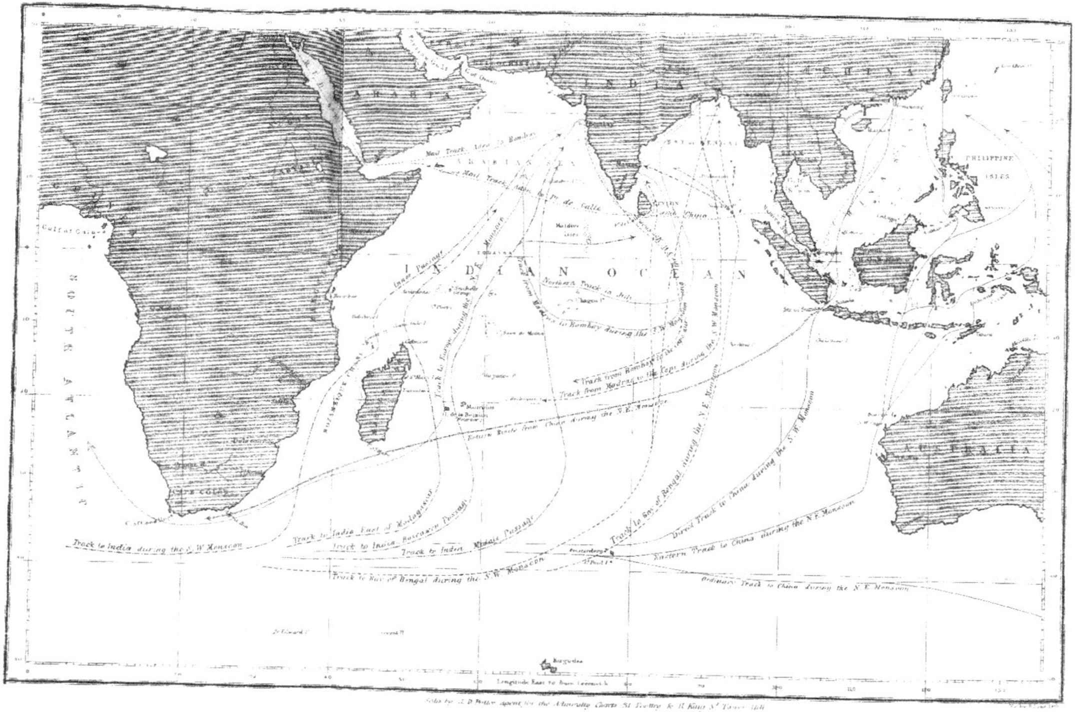

Sold by J. D. Potter agent for the Admiralty Charts 31 Poultry & 11 King St. Tower Hill

Made in the USA
Monee, IL
16 November 2025

35016231R00236